Bonding through Code

Bonding through Code

Theoretical Models for Molecules and Materials

Daniel C. Fredrickson

CRC Press
Taylor & Francis Group
Boca Raton London New York

CRC Press is an imprint of the
Taylor & Francis Group, an **informa** business

MATLAB® is a trademark of The MathWorks, Inc. and is used with permission. The MathWorks does not warrant the accuracy of the text or exercises in this book. This book's use or discussion of MATLAB® software or related products does not constitute endorsement or sponsorship by The MathWorks of a particular pedagogical approach or particular use of the MATLAB® software.

First edition published 2021
by CRC Press
2 Park Square, Milton Park, Abingdon, Oxon, OX14 4RN

and by CRC Press
6000 Broken Sound Parkway NW, Suite 300, Boca Raton, FL 33487-2742

© 2021 Daniel C. Fredrickson

CRC Press is an imprint of Informa UK Limited

British Library Cataloguing-in-Publication Data
A catalogue record for this book is available from the British Library

ISBN: 9781498762212 (hbk)
ISBN: 9780367544874 (ebk)

Typeset in Minion
by Deanta Global Publishing Services, Chennai, India

Dedicated to the memory of
Norris G. Haring, a pioneer in special education
and a role model that I will always look up to.

Contents

Acknowledgments

THIS BOOK HAS EMERGED from the opportunity I have had at the University of Wisconsin–Madison to teach a course that integrates seamlessly with my research interests over many years: *Chem 608—Physical Inorganic Chemistry I: Symmetry and Bonding*. I am deeply thankful to the eight classes of graduate students and advanced undergraduates who have taken on this course, and whose feedback has led me to continuously think about the material and improve the presentation. My colleagues in the Department of Chemistry have also created an environment that is highly supportive of educational innovation. Profs. Thomas Brunold and Clark Landis, both experienced in creating their own versions of Chem 608, have been particularly generous with their time, suggestions, and feedback. Prof. Brunold's Chem 608 notes have had a strong influence on my course that extends to this book. In addition, I am indebted to Prof. Hideo Hosono for graciously hosting me at Tokyo Institute of Technology during my 2016–2017 sabbatical year filled with new experiences, new friends, and the opportunity to share the ideas presented here with a new audience. I am also grateful to Prof. Stephen Lee at Cornell University, who during my first days working with him in graduate school suggested that it might be worth my learning how to plot orbitals with MATLAB®, as well as other mentors whose lessons about bonding, structure, and symmetry helped shape this book, including Profs. Roald Hoffmann, Sven Lidin, and Bart Kahr.

That this work is being seen beyond my classroom owes much to Hilary Lafoe, the book editor at CRC Press who continued to believe in this project even as my attention to it was scattered and offered valuable advice on the shaping the book and bringing it to completion. My thanks also go to Jessica Poile, the editorial assistant who valiantly wrestled with the manuscript pages I sent her—with all of their idiosyncratic formatting and embedded figures—and adapted them for production.

Finally, this book like so many of my projects would not have been possible without the help and encouragement of Dr. Rie Fredrickson, my wife and long-time scientific collaborator, who keeps me focused against the currents of entropy. Rie and our daughters Momoe and Yurie always remind me about what is important in life.

About the Author

Daniel Fredrickson is a Professor in the Department of Chemistry at the University of Wisconsin–Madison, where his research group focuses on understanding and harnessing the structural chemistry of intermetallic phases using a combination of theory and experiment. His interests in crystals, structure and bonding can be traced to his undergraduate research at the University of Washington (B.S. in Biochemistry, 2000) with Prof. Bart Kahr, his Ph.D. studies at Cornell University (2000–2005) with Profs. Stephen Lee and Roald Hoffmann, and his post-doctoral work with Prof. Sven Lidin at Stockholm University (2005–2008). As part of his teaching at UW–Madison since 2009, he has worked to enhance his department's graduate course Physical Inorganic Chemistry I: Symmetry and Bonding, through the incorporation of new material and the development of computer-based exercises.

CHAPTER 1

The Postulates of Quantum Mechanics

CHEMICAL BONDING IS A quantum mechanical phenomenon. As such, we will need familiarity with the properties and behavior of quantum mechanical systems as we work our way from the electronic structure of atoms to increasingly complex molecules and solids. In this first chapter, we will review the fundamental assumptions underlying quantum mechanics (as are covered in more detail in textbooks such as I. N. Levine, *Quantum Chemistry*, Pearson), while simultaneously seeing how they can be realized in the language of MATLAB®, which we will be using to express and explore bonding concepts throughout this book. To illustrate these postulates, we will make use of a common model system: The *particle in a box*.

Postulate 1: *The state of a system is given by its wavefunction. All that can be known about the system is obtainable from this function.*

There are several ways in which this wavefunction can be expressed. When writing equations down on paper, we might express it as a function of the coordinates of the particles in a system: $\psi(x,y,z)$ for a single electron, $\psi(x,y,z,\sigma)$ for an electron with a specific spin, or $\psi(x_1,y_1,z_1,\sigma_1,x_2,y_2,z_2,\sigma_2, ..., x_N,y_N,z_N,\sigma_N)$ for an N-electron system. Alternatively, the wavefunction ψ can be expressed in terms of a series of basis functions, ϕ_n: $\psi = \sum_n c_n\phi_n$, where the state of the system can be completely expressed with the set of coefficients $\{c_n\}$. When using such basis sets, ψ can then be written as a vector of coefficients,

$$\psi = \begin{pmatrix} c_1 \\ c_2 \\ c_3 \\ \vdots \end{pmatrix},$$

and manipulated using matrix math. Finally, a more general notation can be used, which does not refer to any specific mathematical form for the state of the system: The bra-ket notation of Dirac. Here, the state of the system is specified as $|\psi\rangle = \sum_n c_n |\phi_n\rangle$.

Both the functional and vector formats for the wavefunction can be implemented in MATLAB. To see this, let's consider the stationary states for an electron placed in a one-dimensional box of length a, whose potential energy is 0 within the box and ∞ outside of it. As we will see in more detail below, the stationary states of this system are given by the functions

$\psi_n = \left(\dfrac{2}{a}\right)^{\frac{1}{2}} \sin\left(\dfrac{\pi n}{a} x\right)$, where n is a positive integer going over the different

possible states of the system. Such a function can be created in MATLAB by opening the editor, and typing in the following:

```
function psi = particle_in_box_psi(x,a,n)
psi = (2/a)^0.5*sin(pi*n*x/a);
```

Here, the first line serves to define the properties of the function, with "psi" being the output variable, "particle_in_box_psi" being the name of the function, and "x,a,n" being the names of parameters that are given when the function is called. The second line does the actual work of the function, computing the value of psi from the values that are given for x,a,n. Once this function is typed in, it can be saved as the file "particle_in_box_psi.m".

Whenever this file is in your active directory (or a path is given to its location), you can use the function within the MATLAB program. For example, if we were interested in the value of the wavefunction for the $n = 1$ quantum state at 2 Å along the length of a 10 Å box, we would type at the MATLAB command prompt (>>):

```
>> particle_in_box_psi(2,10,1)
```

which returns:

```
ans =
    0.2629
```

Of course, calculating the values of the wavefunction at individual points in space is not a very efficient way of exploring its features. A better way would be to plot the function. To do this, we can make a grid of points along the length of the box by typing:

```
>> x = 0:0.1:10;
```

which defines "x" as a row array of numbers consisting of the values stretching from 0 to 10 in steps of 0.1. The semicolon at the end of the line plays the role of suppressing the screen output of this array, which can prove to be rather distracting. Once this array is defined, you can then plot the wavefunction with the command:

```
>> plot(x,particle_in_box_psi(x,10,1))
```

which should lead to a plot similar to the one shown in Figure 1.1. In this figure, we see that the $n = 1$ wavefunction simply consists of a sine function in which the first half of the wavelength fits exactly in between the walls of the box.

Let's now build on this graphing capability to produce an image summarizing a series of particle-in-a-box states. To do this, we start a new function whose job will be to plot a wavefunction in the context of the box:

```
function psi = particle_in_box_plot(a,n,offset)
x = 0:a/100:a;
psi = particle_in_box_psi(x,a,n);
plot(x,psi+offset,'linewidth',2);
hold on;
plot([0,a],[offset,offset],'linewidth',2,...
'color','black');

% MAKE WALLS
plot([0,0],[0,5],'linewidth',5,'color','black');
plot([a,a],[0,5],'linewidth',5,'color','black');
plot([0,a],[0,0],'linewidth',5,'color','black');
axis off
```

In this function, the first two commands define the psi values for a grid of x points that go from 0 to a in steps of a/100. Then the psi vs. x

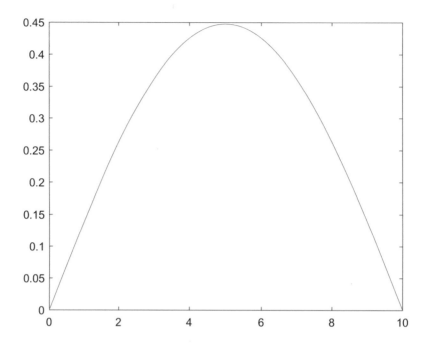

FIGURE 1.1   The $n = 1$ wavefunction for a particle in a box.

is plotted with the y-values shifted by the value of offset, adjusting the height of the function. The command "hold on" tells MATLAB to not erase the current contents of the figure as new elements are added. The next lines draw the $y = 0$ line for the function, and the walls of the box (commands preceded by a "%" symbol are simply comments). Finally, the "axis off" line turns off MATLAB's default axes for the plot.

Typing the command:

```
>> particle_in_box_plot(10,1,1)
```

then leads to the picture in Figure 1.2, and a more complete picture of the series of particle-in-a-box wavefunctions can be built up by repeating this command with different values of n (each with a different value of offset, so that the functions do not overlap):

```
>> figure   % opens new Figure window
>> particle_in_box_plot(10,1,0)
>> particle_in_box_plot(10,2,1.5)
>> particle_in_box_plot(10,3,2*1.5)
>> particle_in_box_plot(10,4,3*1.5)
```

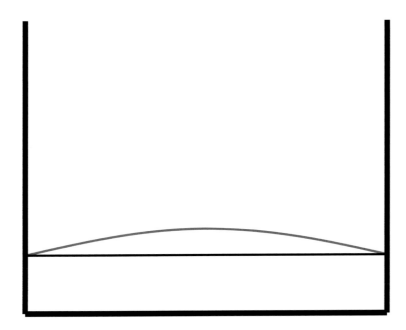

FIGURE 1.2   A better annotated image of the $n = 1$ level for the particle in a box.

The resulting plot (Figure 1.3) shows that with each step in n another node gets added to the wavefunction, corresponding to another half of a wavelength of the sine function being stuffed into the box.

This example illustrates how the wavefunctions can be expressed as functions within MATLAB. Now, let's look at how vector notation is implemented. The functions we have just plotted make up a series that could be used as a basis set for the creation of more complicated functions, i.e.:

$$\psi = \sum_n c_n \phi_n = \sum_n c_n \left(\frac{2}{a}\right)^{\frac{1}{2}} \sin\left(\frac{\pi n}{a} x\right)$$

To create such a function in MATLAB, we can define a vector of coefficients, such as:

```
>> psi_vect = [0.5 0.2 0.2 0.1]'
psi_vect =
    0.5000
    0.2000
    0.2000
    0.1000
```

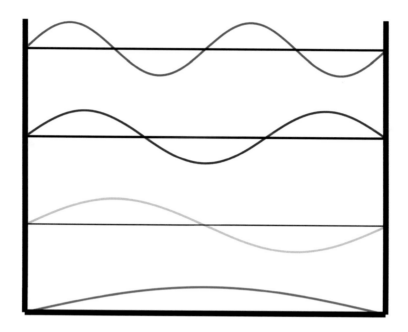

FIGURE 1.3   Levels $n = 1$ through $n = 4$ for the particle in a box.

(where the prime symbol denotes a matrix transpose operation). We then need a MATLAB function for interpreting the vector in terms of basis functions. To make the interpreter, let's save our particle_in_box_plot as a new file with the name particle_in_box_plot_vect.m, and make some changes:

```
function psi = particle_in_box_plot_vect...
(a,coeff,offset)
x = 0:a/100:a;
% New definition of psi
num_components = size(coeff);
num_components = num_components(1);
psi = zeros(size(x));    % array of zeros with same
                         % dimensions as the variable x.
for n = 1:num_components
    psi = psi+coeff(n)*particle_in_box_psi(x,a,n);
end
plot(x,psi+offset);
hold on;
plot([0,a],[offset,offset],'linewidth',2,...
'color',[0,0,0]);
```

```
% MAKE WALLS
plot([0,0],[0,5],'linewidth',5,'color',[0,0,0]);
plot([a,a],[0,5],'linewidth',5,'color',[0,0,0]);
plot([0,a],[0,0],'linewidth',5,'color',[0,0,0]);
axis off
```

Notice that there have been two major changes relative to our earlier function. Instead of using n as an input parameter, we now have "coeff" which is to be a vector of coefficients for the basis functions, each with a different value of n. The lines beginning with "num_components" work out how many coefficients are included in the coeff vector. Then the "for n = 1:num_components ... end" section runs a loop where n goes from 1 to num_components in steps of 1. Within this loop, the line "psi = psi+coeff(n)*particle_in_box_ psi(x,a,n);" adds up the contributions to the total function from the weighted components of basis functions with different values of n.

With this function in place, we now have the ability to plot an endless variety of possible wavefunctions. For example, typing the commands:

```
>> psi_vect = [1.5 -2.8 1.6 0.3]';
>> particle_in_box_plot_vect(10,psi_vect,2*1.5);
```

at the command prompt results in the plot shown in Figure 1.4.

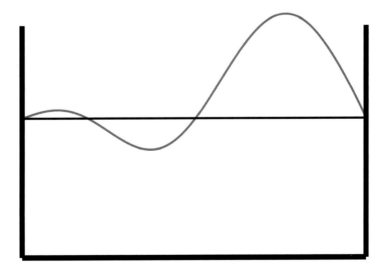

FIGURE 1.4 A wavefunction built from the superposition of different energy eigenfunctions for a particle in a box.

Exercise 1.1. Explain with the help of the MATLAB documentation how the commands "num_components = size(coeff); num_components = num_components(1);" result in num_components containing the number of coefficients in the coeff vector.

Exercise 1.2. Use the MATLAB functions we have created to plot a wavefunction whose $n = 1$, $n = 3$, and $n = 5$ components are all equal to 1, with all other coefficients being zero. The result should look like the function displayed in Figure 1.5.

Postulates 2 and 3: *For every measurable quantity there is an operator in quantum mechanics. The only possible values that can be measured for a property are given by the eigenvalues of its corresponding operator.*

The eigenvalues of an operator are the o_n values that satisfy the condition $\hat{O}\psi_n = o_n\psi_n$, with ψ_n being the eigenfunctions of the operator. Some examples are the operators for a particle's position and momentum along the x-axis: $\hat{x} = x$ and $\hat{p}_x = -i\hbar\dfrac{\partial}{\partial x}$. Another is the Hamiltonian operator, corresponding to the total energy of a system, which for a single particle

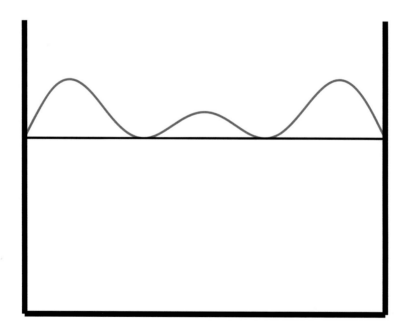

FIGURE 1.5   A wavefunction constructed from equal contributions of the $n = 1$, $n = 3$, and $n = 5$ eigenfunctions.

system is $\hat{H} = -\dfrac{\hbar^2}{2m}\nabla^2 + V(x,y,z)$, where the first term is the kinetic energy operator and the second is the potential energy function.

As we will see below, the eigenfunctions of the Hamiltonian operator are of particular importance as they represent states that are stationary, meaning that (aside from changes in the overall phases of the functions) they do not evolve over time. For the particle in a box, the Hamiltonian operator has a simple form:

$$\hat{H} = -\frac{\hbar^2}{2m}\frac{\partial^2}{\partial x^2} + V(x) \quad \text{with} \quad \begin{cases} V(x) = 0 & \text{when } 0 \le x \le a \\ V(x) = \infty & \text{otherwise} \end{cases}$$

The infinite value of the potential outside the box requires the function to go to zero at the edges of the box. Within the box, the eigenfunctions must follow the equation:

$$\hat{H}\psi_n = -\frac{\hbar^2}{2m}\frac{\partial^2}{\partial x^2}\psi_n = E_n\psi_n$$

The solutions to this can be confirmed to be simply the $\psi_n = \left(\dfrac{2}{a}\right)^{\frac{1}{2}}\sin\left(\dfrac{\pi n}{a}x\right)$ functions we have already implemented in MATLAB code, with the eigenvalues being $E_n = \dfrac{h^2}{8m}\dfrac{n^2}{a^2}$.

This equation for the energies of the particle of the box states contains some features that we often associate with chemical bonding. As the E_n is proportional to n^2, we see that the addition of new nodes to the wavefunction leads to drastic increases in energy. Such is consistent with the idea that shorter wavelengths are connected to higher energy. Also, E_n is inversely proportional to square of the length of the box, so that greater stability is achieved by placing an electron in a larger box. This offers a rationale for why resonance in conjugated systems is often considered to be stabilizing.

Using the offset feature of the particle_in_box_plot function we created earlier, it is possible to better visualize the energy levels of this system. We simply set the offset to be proportional to n^2:

```
>> figure
>> for n = 1:5
particle_in_box_plot(10,n,.5*n^2)
end
```

When you first try this, you will see that the upper function lies above the walls we had specified previously. It is then helpful to adjust the wall heights in the `particle_in_box_plot` function. Once this is done, you can obtain a plot that looks like that given in Figure 1.6.

Postulate 4: *The eigenvectors of a quantum mechanical operator make up a complete set.*

This postulate states that any function ϕ can be expressed as a linear combination of the eigenfunctions of a quantum mechanical operator, ψ_n, of the form: $\phi = \sum_n c_n \psi_n$ or $|\phi\rangle = \sum_n c_n |\psi_n\rangle$. As the eigenfunctions make up an orthonormal set (where $\iiint_{\text{all space}} \psi_n^* \psi_m dV = \langle \psi_n | \psi_m \rangle = \delta_{nm}$), the coefficients in the summation can be easily determined by calculating the overlap of ϕ with each of the basis functions:

$$\iiint_{\text{all space}} \psi_m^* \phi dV = \iiint_{\text{all space}} \psi_m^* \sum_n c_n \psi_n dV = \sum_n c_n \iiint_{\text{all space}} \psi_m^* \psi_n dV = c_m$$

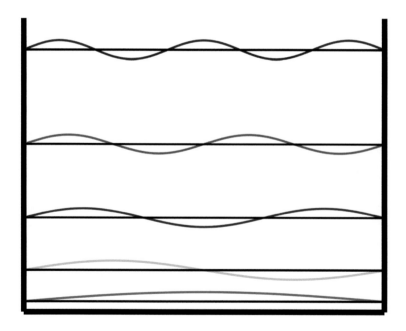

FIGURE 1.6   The eigenfunctions of the particle in a box placed vertically according to their relative energies.

or in bra-ket notation:

$$\left\langle \psi_m \mid \phi \right\rangle = \left\langle \psi_m \left| \sum_n c_n \psi_n \right\rangle \right. = \sum_n c_n \left\langle \psi_m \mid \psi_n \right\rangle = \sum_n c_n \delta_{mn} = c_m$$

As a demonstration of the way in which a set of eigenfunctions can make a complete set, let's consider a delta function, $\delta(x-x_o)$, a function that is defined by the following properties:

$$\delta(x-x_o) = 0 \quad \text{when} \quad x \neq x_o \quad \text{and} \quad \int_{-\infty}^{\infty} \delta(x-x_o) dx = 1.$$

Together, these two properties tell of a function which is zero everywhere but at a single point, where it is so large that the area under that point is 1. Delta functions are of importance in quantum mechanics because they are the eigenfunctions of position:

$$\hat{x}\delta(x-x_o) = x\delta(x-x_o) = x_o\,\delta(x-x_o)$$

The fourth postulate tells us that we should be able to build a delta function using the energy eigenfunctions for the particle in the box. First, we must find the coefficients using the overlap integrals above, as follows:

$$c_n = \left\langle \psi_n \mid \delta(x-x_o) \right\rangle = \int_0^a \left(\frac{2}{a}\right)^{\frac{1}{2}} \sin\left(\frac{\pi n}{a}x\right) \delta(x-x_o) dx = \left(\frac{2}{a}\right)^{\frac{1}{2}} \sin\left(\frac{\pi n}{a}x_o\right)$$

where we have made use of the fact that ψ_n is zero everywhere outside of the box. This equation then gives us a recipe for building up a delta function at any point x_o (as long as it lies inside of the box). Through a modification of the particle_in_box_plot_vect.m this recipe is easy to implement in MATLAB. We first save particle_in_box_plot_vect.m as a new file create_delta.m, and make the changes indicated in black here:

```
function coeff = create_delta(a, x_o, n_max,offset)
for n = 1:n_max
  coeff(n,1) = particle_in_box_psi(x_o,a,n);
end
x = 0:a/1000:a;
```

```
psi = zeros(size(x));    % array of zeros with same
                         % dimensions
                         % as the variable x.
for n = 1:n_max
  psi = psi+coeff(n)*particle_in_box_psi(x,a,n);
end
plot(x,psi+offset);
hold on;
plot([0,a],[offset,offset],'linewidth',2,...
'color',[0,0,0]);

% MAKE WALLS
plot([0,0],[0,6],'linewidth',5,'color',[0,0,0]);
plot([a,a],[0,6],'linewidth',5,'color',[0,0,0]);
plot([0,a],[0,0],'linewidth',5,'color',[0,0,0]);
axis off
```

In this function, the "coeff(n,1)= ..." lines build up an n_max × 1 column vector of coefficients for a delta function centered around the position x_o inside the box. For example, a delta function generated with the first n_max = 40 wavefunctions of the particle in the box centered around x_o = 3 can be created with the commands:

```
>> figure
>> create_delta(10,3,40,2);
>> axis on
```

where the "axis on" command turns the drawing of the coordinate system back on so that the locations of the function's features along the x-axis can be more easily seen. The resulting figure is shown in Figure 1.7. Here, a sharp peak appears centered on $x = 3$, as expected from the properties of our target delta function. In addition, small oscillations occur along the function as it tapers off away from $x = 3$. Such oscillations represent the usual difficulty of representing a sharp discontinuous function with harmonic functions, such as sine and cosine functions.

Exercise 1.3. Demonstrate that our approximation to a delta function becomes increasingly peaked and localized to $x = 3$ as the number n_max is raised. Note that for the approximation to continue to improve, the stepsize in the grid along x should be adjusted so the number of grid points is always significantly higher than the n_max value.

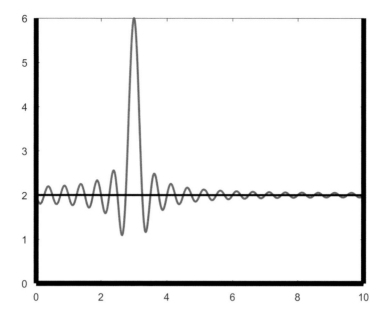

FIGURE 1.7 The reproduction of a deltafunction using a basis set consisting of the lowest 40 particle-in-a-box eigenfunctions.

Exercise 1.4. Use `create_delta.m` as the basis for a new MATLAB function which takes linear combinations of the energy eigenfunctions of the particle in the box to approximate a pair of delta functions: `create_double_delta(a,x_1,x2,n_max,offset)`. Use this function to plot a variety of cases in which the two delta functions are close to or far away from each other.

Postulate 5: *When the state of the system is given by ψ, the average value of a property corresponding to the operator $\hat{O}$ will be the expectation value* $\langle \hat{O} \rangle = \int \phi^* \hat{O} \phi \, d\tau = \langle \phi | \hat{O} | \phi \rangle.$

If we recall that the function ϕ can be written as a linear combination of eigenfunctions of $\hat{O}$, the equation for the expectation value has a simple interpretation:

$$\langle \phi | \breve{O} | \phi \rangle = \left(\sum_m c_m^* \langle \psi_m | \right) \breve{O} \left(\sum_n c_n | \psi_n \rangle \right) = \sum_m \sum_n c_m^* c_n \langle \psi_m | \breve{O} | \psi_n \rangle$$

$$= \sum_m \sum_n c_m^* c_n \langle \psi_m | o_n | \psi_n \rangle = \sum_m \sum_n c_m^* c_n o_n \langle \psi_m | \psi_n \rangle$$

$$= \sum_m \sum_n c_m^* c_n o_n \delta_{mn} = \sum_m |c_m|^2 o_m$$

Here, the expectation value is simply a weighted average over the different eigenvalues o_m of the operator, with the weights $|c_m|^2$ corresponding to the probability of finding the electron in quantum state m.

Postulate 6: *The time-evolution of a system is given by the time-dependent Schrödinger equation:*

$$-\frac{\hbar}{i}\frac{\partial \psi(x,t)}{\partial t} = \hat{H}\psi(x,t)$$

The time-dependent Schrödinger equation is what gives the eigenfunctions of the Hamiltonian operator such an important place in quantum mechanics. For these functions (and assuming that the Hamiltonian itself is time-independent), the above equation reduces to:

$$-\frac{\hbar}{i}\frac{\partial \psi_n(\vec{r},t)}{\partial t} = E_n\psi_n(\vec{r},t)$$

a differential equation that has a straightforward solution:

$$\psi_n(\vec{r},t) = e^{-i(E_n/\hbar)t}\psi_n(\vec{r})$$

The time-dependent eigenfunction is simply the time-independent function multiplied by a phase factor which oscillates with an angular frequency of $E_n/\hbar$. Aside from this phase factor, the function remains unchanged as a function of time, meaning that the eigenfunctions of the Hamiltonian are stationary states.

For other functions, the situation is not nearly so simple. Consider $\phi(\vec{r},t) = \sum_n c_n\psi_n(\vec{r},t)$ which is a linear combination of the eigenfunctions of the Hamiltonian. The time-evolution for this function will then be given by:

$$\phi(\vec{r},t) = \sum_n c_n e^{-i(E_n/\hbar)t}\psi_n(\vec{r})$$

where the phase factors for the different eigenfunctions oscillate at different rates. At different points in time, the various components of the overall function will add together with different phases, with the result being that the shape of ϕ changes with time.

By modifying our previous MATLAB functions, we can make a program to illustrate this effect:

```
function particle_in_box_animate(a,coeff,offset)
x = 0:a/500:a;
num_components = size(coeff);
num_components = num_components(1);
% i h-bar dpsi/dt = E psi
% dpsi/dt = -i E/(h-bar) psi
%
% psi = psi(x,y,z)*exp(-i (E/h-bar)*t)
axis off
axis([0,a,-a,a*2]);
axis manual
hold off
plot([0,0],[0,a*2],'linewidth',5,'color',[0,0,0]);
hold on
plot([a,a],[0,a*2],'linewidth',5,'color',[0,0,0]);
plot([0,a],[0,0],'linewidth',5,'color',[0,0,0]);
plot([0,a],[offset,offset],'linewidth',2,'color',...
  [0,0,0]);

for t = 0:1:400
  psi = zeros(size(x));
  for n = 1:num_components
    E_n = n^2/a^2;
    psi = psi+exp(-i*(E_n)*t)*coeff(n)...
          *particle_in_box_psi(x,a,n);
  end
  h1=plot(x,real(psi)+offset,'color','blue');
  h2=plot(x,imag(psi)+offset,'color','red');
  h3=plot(x,conj(psi).*psi+offset,'linewidth',2,...
     'color','black');

  graph_title = strcat('t = ',num2str(t));
  % Make string for title
  h4=title(graph_title);                % Display title

  pause(.01)
  delete(h1);
  delete(h2);
  delete(h3);
  delete(h4);      % Remove title
end
```

The key steps of the animation are highlighted in black. We begin by setting up a time loop, within which we update the wavefunction psi with the current phase factors for each of its components. As we plot the real and imaginary components of the wavefunction, along with the electron density, we collect handles to the plots in the variables h1, h2, and h3. Having these handles then allows us to delete the functions from the figure, as in "delete(h1)", before we loop to the next time step. The "pause(.01)" command tells MATLAB to pause for 0.01 seconds before continuing to the next command, providing a way for adjusting the frame rate.

Let's start by looking more closely at the behavior of a stationary state. To run an animation of the $n = 1$ state, we type:

```
>> particle_in_box_animate(10,5,5);
```

where we use the coefficient 5 for the $n = 1$ function to scale up the amplitude for visibility. Some snapshots of the evolution of the function are shown in Figure 1.8.

Across this series of plots, the real and imaginary components of the underlying wavefunction change in their relative signs and amplitudes. However, the overall electron density (given by the absolute value of the wavefunction squared, black curve) remains constant.

Exercise 1.5. Show that the $n = 2$ and $n = 3$ wavefunctions of the particle in a box behave in a similar manner.

Now that we have seen how stationary states evolve, let's try a more complicated case, such as a delta function. We can watch how a delta function behaves over time by calculating its coefficients using the create_delta program, opening a new figure, and passing those coefficients on to particle_in_box_animate:

```
>> delta_coeff=create_delta2(10,3,40,2);
>> figure
>> particle_in_box_animate(10,delta_coeff,5);
```

You'll notice that things happen quite rapidly, due to the high-energy (and thus quickly oscillating) functions needed to build up the delta function. It is a good idea to try again after adjusting the time loop in

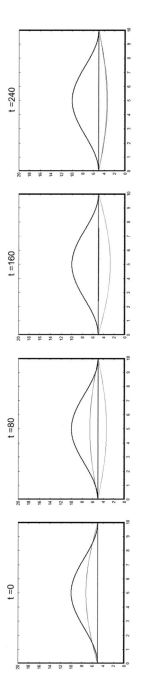

FIGURE 1.8   The time evolution of the real and imaginary components of the $n = 1$ wavefunction of a particle-in-a-box. The probability density (the square of the absolute value of wavefunction) is given as a black curve.

`particle_in_box_animate` to cover a smaller length of time with a smaller sampling, e.g. use "`for t = 0:0.02:5`" instead of "`for t = 0:1:400`".

In Figure 1.9, we show the state of the system at several points between $t = 0$ and $t = 1.0$. At the beginning, the familiar delta function appears just as we generated it with the `delta_coeff` program. Rather quickly, however, the sharp peak of the electron density begins to sink and broaden, and gradually it becomes bimodal at `t = 0.3` and `0.4`, before flattening out into a jumble of small rippling wave crests. This behavior can be connected to the Heisenberg uncertainty principle: At $t = 0$, we know with fairly high accuracy where the particle is. This knowledge about the position implies that we know much less about the momentum of the particle; it could be moving right or left at almost any speed. As time progresses then, the probability distribution for the particle broadens, as its distance from its well-defined starting point at $t = 0$ becomes more and more a mystery.

Exercise 1.6. Demonstrate that δ-like functions generated with a smaller number of particle-in-a-box eigenstates (smaller `n_max`) broaden more slowly, highlighting that decreased precision about a particle's position leads to less uncertainty about its momentum.

Postulate 7: *Following the measurement of an observable, the state will collapse to the eigenfunction corresponding to the eigenvalue measured.*

This postulate is fundamental to defining the relationship between a quantum mechanical system and the person observing it. Observations cannot be made on a system without disturbing it, with the sole exception being the case when a system is already in an eigenstate of the property being measured. Otherwise, to predict the behavior of the system under a measurement, we must first write its wavefunction, ψ, as a linear combination of the eigenfunctions, $\{\phi_n\}$, of the operator corresponding to the property to be measured: $\psi = \sum_n c_n \phi_n$. The probability of measuring the eigenvalue corresponding to state ϕ_n is $|c_n^2|$. In that event, the wavefunction immediately following the measurement will be $\psi' = \phi_n$. It will then evolve according to the time-dependent Schrödinger equation.

Through some work with our current MATLAB code it is possible to simulate this behavior. Let's consider measurements of the particle's position. The probability of finding a particle at point x in the box is

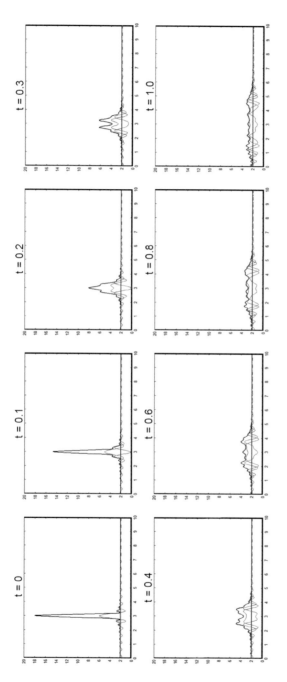

FIGURE 1.9 The time evolution of a delta-like function constructed from the first 40 eigenfunctions of a particle in a box. The high precision of the particles position at $t = 0$ leads to a large uncertainty in its momentum, resulting in a rapid dispersion of the wavefunction. Blue: The real component of the wavefunction. Red: the imaginary component. Black: the square of absolute value of the wavefunction, yielding the probability density.

proportional to $|\psi(x)|^2$. Once we have ψ defined in MATLAB, we can then generate a probability distribution:

```
% Create probability distribution
prob_dist = psi.*psi;
% .* = element-by-element multiplication of
% matrices.
% Normalize probability distribution
prob_dist = prob_dist/sum(prob_dist);
% sum() takes sum of elements in array.
```

With this probability distribution in place that sums up to 1, we can then simulate a measurement by the generation of a random number between 0 and 1 with the command:

```
measurement=rand();
```

Then we check to see where this number lies along the line from $x = 0$ to $x = a$:

```
ngrid_points = size(x);
ngrid_points = ngrid_points(2);

prob_sum=0;

for j = 1:ngrid_points
  if(measurement > prob_sum) && (measurement <= ...
      prob_sum+prob_dist(j))
    x_new = x(j);
  end
  prob_sum=prob_sum+prob_dist(j);
end
```

Here we loop j over the x grid points, and add up the probability values for the points in the variable prob_sum. The if ... end section within this loop runs is activated only if the value of measurement is greater than prob_sum and (&&) less than or equal to (<=) to prob_sum+prob_dist(j). For the grid point along x where these conditions are meant, i.e. the random number falls within its section of the line between 0 and 1, that point is given as the measured value of x: x_new = x(j).

Once the value of x is measured, the wavefunction should collapse to a delta function centered around the measured value. The coefficients for this new function are calculated using the following code pulled and modified from `create_delta.m`:

```
for n = 1:40
    coeff(n,1) = particle_in_box_psi(x_new,a,n);
end
```

which generates coefficients for the first 40 energy eigenfunctions of the particle in a box. From here, we can then continue the animation of `particle_in_box_animate` using these new coefficients, and our simulation of the measurement is complete.

Below is shown the full code for a function that puts all these segments together, `particle_in_box_measure`. It animates the process we just discussed, taking a measurement of x every 2 time units, and showing how the system evolves between these measurements.

```
function particle_in_box_measure(a,coeff,offset)
x = 0:a/500:a;
num_components = size(coeff);
num_components = num_components(1);
axis off
axis([0,a,-a,a*2]);
axis manual
hold off
plot([0,0],[0,a*2],'linewidth',5,'color',[0,0,0]);
hold on
plot([a,a],[0,a*2],'linewidth',5,'color',[0,0,0]);
lot([0,a],[0,0],'linewidth',5,'color',[0,0,0]);
plot([0,a],[offset,offset],'linewidth',...
        2,'color',[0,0,0]);

for measurements = 1:10
  for t = 0:0.02:2
    psi = zeros(size(x));
    for n = 1:num_components
      E_n = n^2/a^2;
      psi = psi+exp(-i*(E_n)*t)*coeff(n)...
            *particle_in_box_psi(x,a,n);
    end
    h1=plot(x,real(psi)+offset,'color','blue');
    h2=plot(x,imag(psi)+offset,'color','red');
```

```
h3=plot(x,conj(psi).*psi+offset,'linewidth',...
    2,'color','black');
graph_title = strcat('\fontsize{20} t = ',...
    num2str(t));
h4=title(graph_title);          % Display title
pause(.01)
delete(h1);
delete(h2);
delete(h3);
delete(h4);  % Remove title
end

prob_dist = psi.*psi;
prob_dist = prob_dist/sum(prob_dist);
measurement=rand();
ngrid_points = size(x);
ngrid_points = ngrid_points(2);
prob_sum=0;
for j = 1:ngrid_points
  if(measurement > prob_sum)&&(measurement <= ...
      prob_sum+prob_dist(j))
            x_new = x(j);
   end
   prob_sum=prob_sum+prob_dist(j);
end
for n = 1:40
    coeff(n,1) = particle_in_box_psi(x_new,a,n);
end
num_components=40;
end
```

Exercise 1.7. Run this simulation and observe its behavior, then annotate the code to particle_in_box_measure, explaining the purpose of each significant command.

CHAPTER 2

Atoms and Atomic Orbitals

INTRODUCTION

In the last chapter, we saw that the measurable energies of a quantum mechanical system and its stationary states are governed by the Schrödinger equation, $\hat{H}\psi_n = E_n\psi_n$. This equation is only solvable analytically for a limited number of systems, with most problems of chemical interest requiring approximant methods. One of the problems that can be solved exactly, however, is the H atom, which provides the starting point for most discussions of atomic structure and bonding in molecules. Here, we will discuss this system in detail. Its eigenstates, the atomic orbitals well-known to all chemists, will then serve as building blocks as we construct bonding schemes for more complicated molecules and materials.

The H atom consists simply of a nucleus, which we can place at the origin of our coordinate system, and an electron at a point that can be described either with the Cartesian coordinates x, y, z or the spherical polar coordinates r, θ, ϕ (with $x = r \sin \theta \cos \phi$, $y = r \sin \theta \sin \phi$, $z = r \cos \theta$). In spherical polar coordinates, the Hamiltonian operator is simply $\hat{H} = -\dfrac{\hbar^2}{2m}\nabla^2 - \dfrac{Ze^2}{r}$, where Z is the number of protons in the nucleus. Here, we see that the potential energy surface that the electron experiences is spherically symmetric around the nucleus. This symmetry (as well as the ability to separate the kinetic energy operator, $-\dfrac{\hbar^2}{2m}\nabla^2$, into radial and angular terms) allows us to write each solution to the Schrödinger equation as the product of a radial function and an angular function: $\psi = R(r)Y(\theta,\phi)$.

As the electron is highly attracted to the nucleus, the probability of finding the electron far from it, $\propto r^2 |\psi|^2$, goes to zero as $r \to \infty$. This boundary condition leads to the quantization of the energy levels, giving each energy eigenfunction three quantum numbers such that:

$$\hat{H}\psi_{n,l,m} = E_n\psi_{n,l,m} \text{ with } \psi_{n,l,m} = R_{n,l}(r)Y_{l,m}(\theta,\phi)$$

where the radial wavefunction $R_{n,l}(r)$ is the product of a polynomial and an exponential decay function, and the angular wavefunctions $Y_{l,m}(\theta,\phi)$ are the classic spherical harmonics.

The quantum numbers n, l, and m lay out the shell structure of the atom; n is the principal quantum number, which (for a 1-electron atom) solely determines the energy of an electron with this wavefunction. For each n shell, there are various possible values for l, the azimuthal quantum number, with allowed values of l being integers in the range $l = 0, 1, 2, ..., n - 1$. The l values 0, 1, 2, and 3 are historically associated with the letters s, p, d, and f, respectively. Finally, for each value of l, a range of magnetic quantum numbers are allowed, $m = 0, \pm1, \pm2, ..., \pm l$. This set of spatial wavefunctions describes the atomic orbitals, each of which can contain two electrons (one up-spin, one down-spin).

THE RADIAL WAVEFUNCTION

The radial wavefunctions $R_{n,l}(r)$ consist of a product of an exponential decay function tending to concentrate the wavefunctions around the nucleus and a polynomial that introduces oscillations in the sign of the wavefunction to provide orthogonality to other wavefunctions:

$$R_{n,l}(r) = N\left(\frac{2Zr}{n}\right)^l L_{n-l-1}^{2l+1}\left(\frac{2Zr}{n}\right)e^{-\frac{Zr}{n}}$$

where r is in atomic units, N is a normalization constant, and $L_{n-l-1}^{2l+1}(x)$ refers to the generalized Laguerre polynomials. While this function has a rather complicated form, it is rather easy to implement in MATLAB® (at least for the 2014b version and above). To do this, we create a new function, R_n_l.m:

```
function psi_r = R_n_l(r, n, l, Z)

    % Calculate value of radial wavefunction
    N = ((2*Z/n)^3*factorial(n-l-1)/(2*n...
        *factorial(n+l)))^0.5;
    psi_r = N*(2*Z*r/n).^(l)*laguerreL(n-l-1,...
            2*l+1, 2*Z*r/n).*exp(-Z*r/n);
```

With this function in place, we can make plots in MATLAB of the radial dependence of radial function. For instance, say we wanted to plot a 1s orbital of an H atom ($Z = 1$) over the range of 0.0 a_o to 5.0 a_o in steps of 0.01 a_o, we could do it in the following way:

```
≫ r = 0.0: 0.01 : 5.0;
% define r as array from 0 to 5 in steps of 0.01.
≫ size(r)   % determine number of points in array r

ans =

      1   501   % Answer = r is an array with one
                % row of 501 values.

≫ for j = 1:501
       Rfunction(j) = R_n_l(r(j), 1, 0, 1);
   end
≫ plot(r, Rfunction);
```

The result is shown in Figure 2.1: An exponential decay of the 1s orbital with increasing distance from the nucleus.

We now take the above commands and make them into a reusable function for plotting radial wavefunctions:

```
function plot_R_n_l(n,l,Z,r_min,r_max,color_variable)

r = r_min: 0.01 : r_max;
npoints = size(r);    % determine number of points
                      % in array r
npoints = npoints(2); % keep only 2nd number of
                      % npoints
```

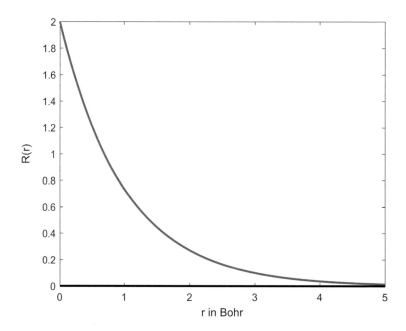

FIGURE 2.1   The radial wavefunction for the 1s orbital of H.

```
for j = 1:npoints
    Rfunction(j) = R_n_l(r(j), n, l,Z);
end

plot(r, Rfunction,'color', color_variable);
xlabel('r in Bohr');   % label for x-axis
ylabel('R(r)');       % label for y-axis
hold on
plot([r_min r_max],[0,0],'color','black',...
        'linewidth',2);
```

The 1s function can then be simply plotted with:

```
>> plot_R_n_l(1,0,1, 0.0, 5.0, 'blue');
>> title('\fontsize{20} 1s')
```

Exercise 2.1. Use the plot_R_n_l function to plot the radial wavefunctions of the 1s, 2s, 3s, 4s, 2p, 3p, 3d, and 4d orbitals. From these plots verify that the number of radial nodes is equal to $n - 1 - l$. Examples are shown in Figure 2.2.

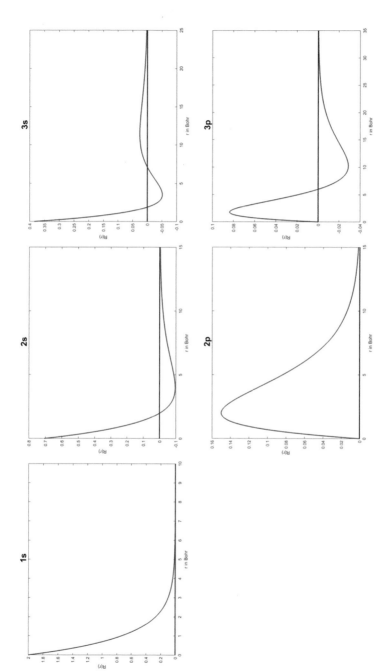
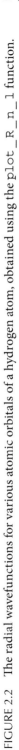

FIGURE 2.2   The radial wavefunctions for various atomic orbitals of a hydrogen atom, obtained using the plot_R_n_l function.

Exercise 2.2. Make a modified version of the plot_R_n_1.m func-
tion to plot the radial probability density for the electron, $4\pi r^2|R(r)|^2$, and
reproduce the plots shown in Figure 2.3.

VISUALIZING ATOMIC ORBITALS WITH MATLAB: THE ANGULAR WAVEFUNCTIONS

Now that we've developed a feel for the radial component of atomic orbitals, let's turn to their angular functions, $Y(\theta, \phi)$. A simple way to graph
these functions is through radial plots, surfaces drawn around the nuclear
position on which the distance from the surface to the nucleus is proportional to the magnitude of the angular function along that direction.

The first step in creating such a plot is to generate a grid of points upon
the surface of a sphere, each of which can be moved toward or away from
the nucleus to reflect the shape of the angular wavefunction. This can be
done with the MATLAB function sphere:

```
≫ [x,y,z] = sphere(40);  % obtain array of points
                         % for sphere with fineness
                         % given by 40
≫ surf(x,y,z)       % draw surface through points
                    % given by arrays x,y,z
≫ axis equal        % set all axes to same scale
≫ light             % turn on the lights
```

After typing in these lines, you should obtain an image similar to that
in Figure 2.4.

Next, the points obtained can be converted to spherical polar coordinates (r, θ, ϕ), so that the r values can be adjusted according to the value
of $Y(\theta, \phi)$:

```
≫   [phi, theta_matlab, r] = cart2sph(x,y,z);
≫ theta = pi/2 - theta_matlab;
% Matlab defines theta as the angle out of the xy
% plane toward the z-axis, rather than the angle
% away from the z-axis (the convention in physics/
% chemistry).
```

Now we can set the values in the r array to equal the values of $Y(\theta, \phi)$
at each point on the sphere, convert the point positions back to Cartesian

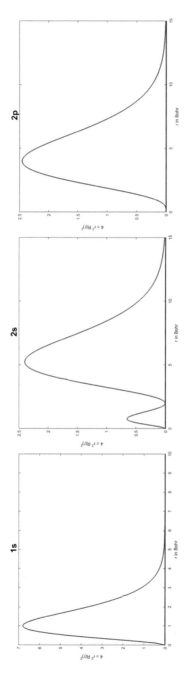

FIGURE 2.3   Radial probability distributions for the H 1s, 2s, and 2p atomic orbitals.

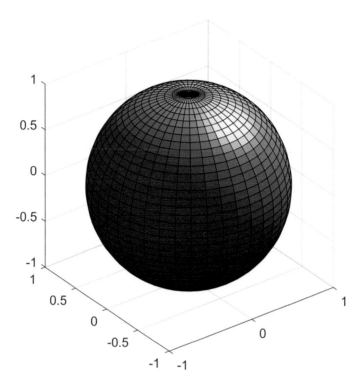

FIGURE 2.4 Plot obtained using MATLAB's built-in `sphere` and `surf` functions.

coordinates, and then plot the surface. Here's how we might start to do this for a p$_z$ orbital, where $Y(\theta, \phi) \propto cos(\theta)$:

```
≫  r_pz = cos(theta);
     % calculates cosine of all elements of theta array
≫  [xnew,ynew,znew] = sph2cart(phi,theta_matlab,...
                       abs(r_pz));
   % converts back to Cartesian coordinates.
   % Note that we use "theta_matlab" as we are using
   % the MATLAB function for transforming between
   % spherical and Cartesian coordinates.
≫ surf(xnew,ynew,znew);
≫ axis equal;
≫ light;
```

The result is shown in Figure 2.5. In the above commands, we used only the absolute values of $Y(\theta, \phi)$ to plot the surface, as the radial plotting scheme makes it difficult to use distances to represent negative values. We

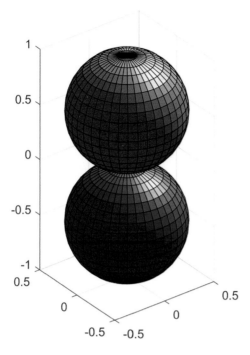

FIGURE 2.5   Preliminary image of the angular component of a p$_z$ orbital.

can get around this by using color to represent the signs of the function. We accomplish this by building a new array that contains a color value for each point on our surface, depending on whether $Y(\theta, \phi)$ is positive or negative. Here it is helpful to know that colors can be specified in MATLAB (as well as other programs, such as Microsoft Word and Adobe Illustrator) through RGB (red-green-blue) color codes:

red = [255 0 0]/255   light green = [0 255 0]/255
blue = [0 0 255]/255   yellow = [255 255 0]/255
black = [0 0 0]   white = [255 255 255]/255

All we need to do then is define a color array with three numbers for each point in the r_pz array, and then populate these numbers with different colors depending on the size of $Y(\theta, \phi)$:

```
>> npoints = size(r)

ans =

    41   41
```

```
≫ color_values = zeros(npoints(1),npoints(2),3);
≫ for j = 1:npoints(1)
   for k = 1:npoints(2)
     if(r_pz(j,k) > 0)
       cdata(j,k,:) = [0 0 0];
       % color data = black for positive
     end
     if(r_pz(j,k) < 0)
       cdata(j,k,:)=[255 255 255]/255;
       % color data = white for negative
     end
   end
  end
≫ surf(xnew,ynew,znew,'CData',cdata,'EdgeColor',...
  'none');
≫ axis equal; light;
```

Entering these commands yields the plot of the p_z orbital shown in Figure 2.6.

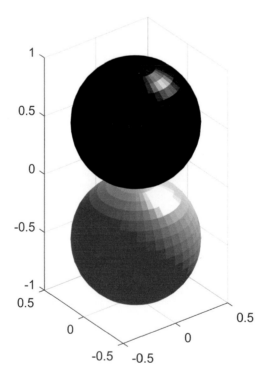

FIGURE 2.6 Plot of the angular component of the p_z orbital, with black and white denoting opposite signs of the function.

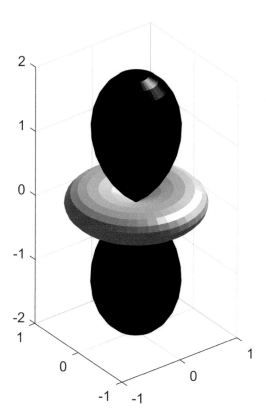

FIGURE 2.7 The angular part of the d$_{z2}$ orbital.

Another example is the d$_{z2}$ orbital: $Y(\theta, \phi) = N (3cos^2(\theta) - 1)$, the code for which is shown below, with the resulting plot illustrated in Figure 2.7.

```
>>    r_dz2 = 3*cos(theta).*cos(theta)-1;
   % Important note: ".*" means that each of
   % element of an array is multiplied by its
   % corresponding element in the next array, as
   % opposed to usual matrix multiplication. This is
   % essential whenever evaluating a function over an
   % array of points.
>> [x2,y2,z2] = sph2cart(phi,theta_matlab,abs(r_dz2));
>>  npoints = size(r)
>> cdata = zeros(npoints(1),npoints(2),3);
>> for j = 1:npoints(1)
   for k = 1:npoints(2)
     if(r_dz2(j,k) > 0)
```

```
        cdata(j,k,:) = [0 0 0];
        % black for positive
      end
      if(r_dz2(j,k) < 0)
        cdata(j,k,:) = [255 255 255]/255;
        % white for negative
      end
    end
  end
≫ surf(x2,y2,z2,'CData',cdata,'EdgeColor','none');
≫ axis equal; light;
```

At this point, it becomes clear that a large series of instructions will need to be repeated every time we plot a new orbital. The process can be simplified by incorporating these steps into a MATLAB function that can be run any time we want to draw an orbital. The outline for such a function is given below:

```
function draw_atomic_orbital(s,px,py,pz,dz2,...
dx2y2,dxy,dxz,dyz)
% s,px,...,dyz = coefficients of corresponding
% orbital in wavefunction.

[x,y,z] = sphere(40);
[phi, theta_matlab, r] = cart2sph(x,y,z);
theta = pi/2 - theta_matlab;

r_s = 1/(2*pi^0.5);
r_px = (3/(4*pi))^0.5.*sin(theta).*cos(phi);
r_py = ...
r_pz = ...
r_dz2 = ...
r_dx2y2 = ...
r_dxy = ...
r_dxz = ...
r_dyz = ...

r_new = s*r_s + px*r_px + py*r_py + pz*r_pz +...
          dz2*r_dz2 + dx2y2*r_dx2y2 + ...
          dxy*r_dxy + dxz*r_dxz + dyz*r_dyz;
[x_new, y_new, z_new] = sph2cart(phi,...
                        theta_matlab,abs(r_new));
```

```
npoints = size(r);
color_values = zeros(npoints(1),npoints(2),3);
for j = 1:npoints(1)
   for k = 1:npoints(2)
      if(r_new(j,k) > 0)
         color_values(j,k,:) = [0 0 0];
         % black for positive
      end
      if(r_new(j,k) < 0)
         color_values(j,k,:) = [255 255 255]/255;
         % white for negative
      end
   end
end
surf(x_new,y_new,z_new,'CData',color_values,...
      'EdgeColor','none');
axis equal; light; % See upper right for result.
```

Exercise 2.3. Fill in the blanks in the function above indicated by "= ..."
with the normalized angular wavefunctions for the atomic orbitals, as
may be found on Wikipedia.

Exercise 2.4. Test the completed function by making plots of all of the s, p,
and d orbitals.

Exercise 2.5. For everything we have done so far, we have assumed that we
are working with an atom centered at the origin. This will change when
we start drawing molecular orbitals, where at most one atom will be at the
origin. Copy the draw_atomic_orbital.m function file to draw_
atomic_orbital_xyz.m, and modify the function to plot the orbital
at a position (x, y, z) given to the function as input variables.

COMBINING THE RADIAL AND ANGULAR FUNCTIONS

Now that we have analyzed the radial and angular parts of the H atom
wavefunctions separately, we are in a position to view the full functions
in 3D space. Representing these functions becomes a little challenging, as
we are accustomed to visualizing objects in three dimensions, but plotting the orbitals as continuous functions of x, y, and z would require four
dimensions. To overcome this, we can use isosurfaces—surfaces that correspond to a single value of the function—for a series of values to get a
sense of the spatial distributions of the wavefunctions.

The first step is to evaluate the values of $\psi_{n,l,m} = R_{n,l}(r)Y_{l,m}(\theta,\phi)$ at specific points in space:

```
function psi = H_atom_psi(x,y,z,n,c_s,c_px,...
                c_py,c_pz)
% We'll limit ourselves to just the s and p
% orbitals for now
% with Z = 1.

[phi, theta_matlab, r] = cart2sph(x,y,z);
theta = pi/2 - theta_matlab;
R_s = R_n_l(r, n, 0, 1);
R_p = 0;
if (n > 1)
    R_p = R_n_l(r, n, 1, 1);
end
Y_s = 1/(2*pi^0.5);
Y_px = (3/(4*pi))^0.5.*sin(theta).*cos(phi);
Y_py = (3/(4*pi))^0.5.*sin(theta).*sin(phi);
Y_pz = (3/(4*pi))^0.5.*cos(theta);

psi = c_s*R_s*Y_s + c_px*R_p*Y_px + ...
      c_py*R_p*Y_py + c_pz*R_p*Y_pz;
```

Next, we start a new file and evaluate the above function over a grid of points, keeping track of the x, y, and z values for each grid point:

```
x_j = 0; % counter for grid points along x
y_j = 0; % counter for grid points along y
z_j = 0; % counter for grid points along z
for x = -10:.1:10
    x_j = x_j+1;
    y_j=0;
    for y = -10:.1:10
        y_j = y_j+1;
        z_j = 0;
        for z = -10:.1:10
            z_j = z_j+1;
            X_grid(x_j,y_j,z_j) = x;
            Y_grid(x_j,y_j,z_j) = y;
            Z_grid(x_j,y_j,z_j) = z;
        end
```

```
      end
end
psi_grid = H_atom_psi(X_grid,Y_grid,Z_grid,n,c_s,...
            c_px,c_py,c_pz);
% n, c_s, c_px, c_py, c_pz still need to be
% specified
```

Then we can use the isosurface and patch commands in MATLAB to plot surfaces of the function at different values.

```
psi_value = .1
psi_surface = isosurface(X_grid,Y_grid, Z_grid,...
        psi_grid, psi_value)
patch(psi_surface,'FaceColor',[0 0 0], ...
        'EdgeColor','none','FaceAlpha',.8);
% FaceAlpha sets transparency of surface

hold on

psi_surface = isosurface(X_grid,Y_grid, Z_grid,...
        psi_grid, -psi_value)
patch(psi_surface,'FaceColor',[.9 .9 .9],...
        'EdgeColor','none','FaceAlpha',.8);
```

This code can then be combined together into a single function for plotting isosurfaces:

```
function isosurface_H_atom(n,c_s,c_px,c_py,c_pz,...
    psi_value)

x_j = 0;
y_j = 0;
z_j = 0;
for x = -20:.4:20 % Some adjustments to the grid
            % spacing and range may be necessary for
            % each orbital, and the speed of your
            % computer. The longer x range here is set
            % for plotting a px orbital.
    x_j = x_j+1;
    y_j = 0;
```

```
    for y = -10:.4:10
      y_j = y_j+1;
      z_j = 0;
      for z = -10:.4:10
        z_j = z_j+1;
        X_grid(x_j,y_j,z_j) = x;
        Y_grid(x_j,y_j,z_j) = y;
        Z_grid(x_j,y_j,z_j) = z;
      end
    end
  end

  psi_grid = H_atom_psi(X_grid,Y_grid,Z_grid,n,...
        c_s,c_px,c_py,c_pz);

  psi_surface = isosurface(X_grid,Y_grid, Z_grid,...
        psi_grid, psi_value)

  % Printing out the max and min of psi can help set
  % an isosurface value.
  psi_max = max(max(max(psi_grid)))
  psi_min = min(min(min(psi_grid)))

  patch(psi_surface,'FaceColor',[0 0 0],...
        'EdgeColor','none','FaceAlpha',.8);
  % FaceAlpha sets transparency of surface

  hold on
  psi_surface = isosurface(X_grid,Y_grid, Z_grid,...
        psi_grid, -psi_value)
  patch(psi_surface,'FaceColor',[.9 .9 .9],...
        'EdgeColor','none','FaceAlpha',.8);
  axis equal
  light
```

Using this function, a plot such as that for a 3p orbital in Figure 2.8 can be obtained.

FOCUSING ON THE VALENCE ELECTRONS: SLATER-TYPE ORBITALS

In discussions of chemical bonding, the details of the exact functions for the atomic orbitals are not all necessary for two reasons. First, we

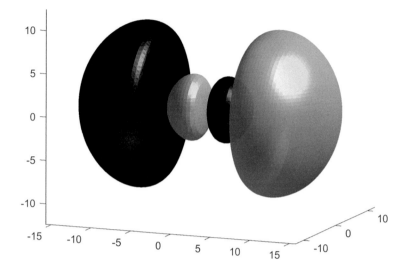

FIGURE 2.8   Isosurface plot of a H 3p orbital. Isosurface value = 0.008.

generally focus on the atomic orbitals containing the valence electrons, as we consider the electrons in lower energy shells to be core-like and chemically inert. Also, we envision the interactions between atoms to occur as the outer regions of their atomic orbitals overlap. The actual form of the wavefunction near the nucleus is not really essential for such purposes; the main role of the complicated shapes of the wavefunctions close to the core is to maintain orthogonality to lower energy atomic orbitals. When we restrict our attention to the long-range character of the valence orbitals, it becomes convenient to use simplified functions known as Slater-type orbitals. These functions have the form $\psi_{n,l,m}(r,\theta,\phi) = N\, r^{n-1}e^{-\zeta r}\, Y_{l,m}(\theta,\phi)$, where N is a normalization constant equal to $(2\zeta)^{n+\frac{1}{2}}/((2n)!)^{\frac{1}{2}}$. Using a single Slater-type orbital to represent each valence orbital in a bonding problem is an example of the use of a *minimal basis set*. Let's look in more detail at the character of these functions.

Exercise 2.6. Write MATLAB functions for plotting the radial wavefunction and radial probability distributions for Slater-type orbitals. Generate plots showing how $R(r)$ and the radial probability density as functions or r vary with changes in n and ζ, as shown in Figure 2.9. In what ways do these parameters change the shape of the electron density distribution associated with the orbital?

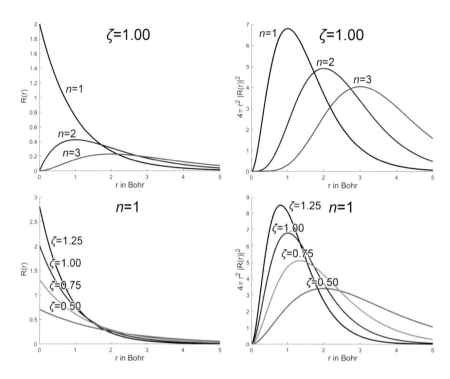

FIGURE 2.9 Examples of radial wavefunction and probability distribution curves for Slater-type orbitals with different values of n and ζ.

Exercise 2.7. In the plots you obtained in Exercise 2.6, why do the magnitudes of the features along the y-axis gradually become smaller as n increases or ζ decreases?

CHAPTER 3

Overlap between Atomic Orbitals

INTRODUCTION

In the previous chapter, we created functions for modeling the atomic orbitals we will use to build up bonding schemes for molecules and solids. As we bring atoms together to form more complex systems, a major factor in their interactions will be the extent to which the atomic orbitals on neighboring atoms overlap with each other. This overlap between two such orbitals, say ϕ_i and ϕ_j, is given by the overlap integral $S_{ij} = \iiint_{\text{all space}} \phi_i^* \phi_j d\tau$.

Here, we will see how these integrals can be evaluated using MATLAB® and examine some of their properties.

PARAMETERS FOR SLATER-TYPE ORBITALS

For the following discussion, we will use C and H atoms in our examples. The parameters we use for these atoms are taken from a Hückel model of benzene parameterized to fit the results of DFT calculations.[1,2] The H 1s orbital has a ζ value of $\zeta_s = 1.46\ a_o^{-1}$. The C parameters are $\zeta_s = 2.25\ a_o^{-1}$ for the 2s orbital and $\zeta_p = 2.22\ a_o^{-1}$ for the 2p orbitals. Later on, we will also need ionization energies (H_{ii}'s) of these orbitals, which were also calibrated against the DFT results. These values are –7.53 eV, –11.24 eV, and –7.86 eV for the H 1s, C 2s, and C 2p orbitals, respectively.

These parameters can be saved in arrays for future use in our MATLAB work as follows:

```
>> %          n_s  H_ii(s) zeta_s n_p  H_ii(p) zeta_p
>> H_parameters = [ 1   -7.53  1.46  0   0.0    0.0 ];
>> C_parameters = [ 2  -11.24  2.25  2  -7.86   2.22];
```

We'll write our MATLAB function with this array format in mind so that the parameters can be easily provided and used. Analogous arrays can of course be defined for other elements as well. In future chapters, we will add to this list.

COMBINING THE RADIAL AND ANGULAR FUNCTIONS

Our first step in calculating the overlap integrals between atomic orbitals is to develop a function for evaluating values of an orbital at specific points in space. Consider an atom at the position within a molecule is given by the Cartesian coordinates atom_pos = [x_atom y_atom z_atom]. The value of an atomic orbital at the point [x y z] will then depend on that point's position relative to the nucleus:

```
% measure position of point relative to nucleus
x_rel = x - atom_pos(1);
y_rel = y - atom_pos(2);
z_rel = z - atom_pos(3);
```

We can then evaluate the usual function $\psi = R(r)Y(\theta, \phi)$ where r, θ, and ϕ represent x_rel, y_rel, z_rel in spherical polar coordinates:

```
[phi, theta_matlab, r] = ...
cart2sph(x_rel,y_rel,z_rel);
theta = pi/2 - theta_matlab;
  % Transform theta to conventional definition
```

After this transformation is done, the radial and angular components of the wavefunction can be calculated using code we developed in the last chapter. The completed function is shown below:

```
function psi = STO_psi(x,y,z,atom_pos,...
STO_parameters, coefficients)
%
%  atom_pos = [x_nucleus, y_nucleus, z_nucleus]
```

```matlab
%  STO_parameters = [n_s Hii_s zeta_s  n_p
%  Hii_p zeta_p ]
%  coefficients = [c_s c_px c_py c_pz]
%

% measure position of point relative to nucleus
x_rel = x - atom_pos(1);
y_rel = y - atom_pos(2);
z_rel = z - atom_pos(3);

% Extract orbitals
c_s = coefficients(1);
c_px = coefficients(2);
c_py = coefficients(3);
c_pz = coefficients(4);

n_s   = STO_parameters(1);
Hii_s = STO_parameters(2);
zeta_s = STO_parameters(3);
n_p   = STO_parameters(4);
Hii_p = STO_parameters(5);
zeta_p = STO_parameters(6);


[phi, theta_matlab, r] = ...
cart2sph(x_rel,y_rel,z_rel);
theta = pi/2 - theta_matlab;

% Define normalization constant, N
N_s = (2*zeta_s)^(n_s+0.5)/(factorial(2*n_s))^0.5;

% Calculate value of radial wavefunction
R_s = N_s*r.^(n_s-1).*exp(-zeta_s*r);

% Calculate value of angular wavefunction
Y_s = 1/(2*pi^0.5);
R_p = 0; Y_px = 0; Y_py = 0; Y_pz = 0;

if(n_p > 0)
    % repeat process for p orbital if defined.
    N_p = (2*zeta_p)^(n_p+0.5)/(factorial(2*n_p))^0.5;
    R_p = N_p*r.^(n_p-1).*exp(-zeta_p*r);
    Y_px = (3/(4*pi))^0.5.*sin(theta).*cos(phi);
```

```
        Y_py = (3/(4*pi))^0.5.*sin(theta).*sin(phi);
        Y_pz = (3/(4*pi))^0.5.*cos(theta);
    end

    psi = c_s*R_s.*Y_s + c_px*R_p.*Y_px + c_py*R_p.*...
          Y_py + c_pz*R_p.*Y_pz;
```

VISUALIZING ISOSURFACES OF SLATER-TYPE ORBITALS

With this function in place, it becomes possible to visualize the radial and angular functions of the Slater-type orbitals together through isosurfaces, as we did previously for the full atomic orbitals. To do this, we first evaluate any given atomic orbital over a grid of points, keeping track of the x, y, and z values for each grid point:

```
% Set up grid in Cartesian space
x_j = 0;  % x_j, y_j, z_j = counters for keeping
            % track of grid points
y_j = 0;
z_j = 0;
for x = -5:.1:5
  x_j = x_j+1;
  y_j=0;
  for y = -5:.1:5
    y_j = y_j+1;
    z_j = 0;
    for z = -5:.1:5
      z_j = z_j+1;
      X_grid(x_j,y_j,z_j) = x;
      Y_grid(x_j,y_j,z_j) = y;
      Z_grid(x_j,y_j,z_j) = z;
    end
  end
end

% Evaluate function over grid of points
psi_grid = STO_psi(X_grid,Y_grid,Z_grid,...
          atom_pos, STO_parameters, coefficients)
    % Still need to define atom_pos,
    % STO_parameters, coefficients.
```

Then we can use the isosurface and patch (plotting) commands in MATLAB to plot surfaces of the function at different values.

```
psi_value = .1
psi_surface = isosurface(X_grid,Y_grid, Z_grid,...
        psi_grid, psi_value)
patch(psi_surface, 'FaceColor',[0 90 255]/255,...
        'FaceAlpha',.8);
% FaceAlpha sets transparency of surface

hold on

psi_surface = isosurface(X_grid,Y_grid, ...
        Z_grid, psi_grid, -psi_value)
patch(psi_surface, 'FaceColor',[0 140/255 0],...
        'FaceAlpha',.8);
```

To implement this procedure, we simply combine all the above code into a single function for plotting isosurfaces:

```
function isosurface_STO_orbital(atom_pos, ...
    STO_parameters,coefficients, psi_value)

% Set up grid in Cartesian space
x_j = 0;
y_j = 0;
z_j = 0;
for x = -5:.1:5
  x_j = x_j+1;
  y_j=0;
  for y = -5:.1:5
    y_j = y_j+1;
    z_j = 0;
    for z = -5:.1:5
      z_j = z_j+1;
      X_grid(x_j,y_j,z_j) = x;
      Y_grid(x_j,y_j,z_j) = y;
      Z_grid(x_j,y_j,z_j) = z;
    end
  end
end

% Evaluate function over grid of points
psi_grid = STO_psi(X_grid,Y_grid,Z_grid,...
        atom_pos, STO_parameters, coefficients);
```

```
% Draw positive isosurface
psi_surface = isosurface(X_grid,Y_grid, Z_grid,...
        psi_grid, psi_value);
patch(psi_surface,'FaceColor',[0 90 255]/255,'...
        FaceAlpha',.8,'EdgeColor','none');
% FaceAlpha sets transparency of surface:...
% 1 = solid, 0 = invisible.

hold on

% Draw negative isosurface
psi_surface = isosurface(X_grid,Y_grid, Z_grid,...
        psi_grid, -psi_value);
patch(psi_surface,'FaceColor',[0 140/255 0],...
        'FaceAlpha',.8,'EdgeColor','none');

axis equal  % set scales along x, y, and z to be equal
axis off
light       % turn on light
```

As example of using this function, let's plot the isosurface for a C 2p orbital at value of 0.2:

```
>> C_parameters = [2 -11.24 2.25 2 -7.857 2.22];
>> isosurface_STO_orbital([0 0 0], C_parameters,...
   [0 1 0 0], .2);
```

The result of running these commands is the surface along which a p_x orbital is equal to ± 0.2 (see Figure 3.1, left). The positive surface is plotted in blue, while the negative surface is shown in green. The colors used and the transparencies of the surfaces can be adjusted by changing the values for FaceColor and FaceAlpha in the patch commands in the isosurface_STO_orbital function. By playing with the transparency level, and overlaying isosurfaces for several values at once, visualization of the radial dependence also becomes possible (see Figure 3.1, right). These images provide confidence that we have correctly represented the atomic orbitals as functions of x, y, and z.

Exercise 3.1. Use the isosurface_STO_orbital function to make concentric isosurfaces of the C valence orbitals at values from 0.4 to 0.05 in steps of 0.5. Hint: Using for loops can make this repetition simpler. For

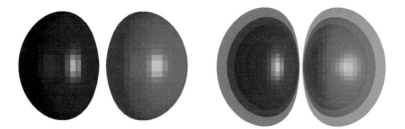

FIGURE 3.1   A Slater-type orbital model of a C 2p atomic orbital, shown with a single isosurface (left), and with several partially transparent isosurfaces (right).

this to work optimally, it is a good idea to remove the `light` command from the function, and apply it after all plots have been made.

PROGRAMMING OVERLAP INTEGRALS IN MATLAB

Now that we have a function that can evaluate the values of an atomic orbital at an array of points in Cartesian space, we move to the task of calculating overlap integrals between them. For such integrals involving products of Slater-type orbitals, there are in fact analytical solutions. However, these are of a complicated form and working through them would not contribute much to our understanding of bonding interactions. Instead, it is more convenient to use MATLAB's numerical algorithms for carrying out definite integrals.

Let's say that we wanted to calculate the overlap between two orbitals, ψ_1 and ψ_2, which are centered at points [x1 y1 z1] and [x2 y2 z2] with atomic orbital coefficients and Slater-type orbital parameters given by `STO_parameters1`, `coeff1`, `STO_parameters2`, and `coeff2`. To calculate the integral of their product over all space, we first make versions of these functions that are dependent only on x, y, and z, with the other input parameters of `STO_psi` being treated as constants:

```
psi1 = @(x,y,z) STO_psi(x,y,z,[x1 y1 z1],...
STO_parameters1, coeff1);
psi2 = @(x,y,z) STO_psi(x,y,z,[x2 y2 z2],...
STO_parameters2, coeff2);
```

where the "@(x,y,z)" prefix indicates that rather than variables with a single value, psi1 and psi2 are being defined as functions of x, y, and z. For example, if we use the C $2p_x$ orbital from the last section, psi1 could be defined as:

```
>> psi1 = @(x,y,z) STO_psi(x,y,z,[0 0 0],...
           C_parameters, [0 1 0 0]);
```

After such a definition, `psi1` can be used in MATLAB just as any other function. For example, the following code leads to the radial plot shown in Figure 3.2.

```
>> figure
>> plot(0:.1:5, psi1(0:.1:5,0,0))
```

Once these abbreviated functions are made, we are able to calculate integrals involving them with the `integral3` (triple integral) function:

```
Sintegrand = @(x,y,z) psi1(x,y,z).*psi2(x,y,z);
S_ij = integral3(Sintegrand,-Inf,Inf,-Inf,Inf,...
          -Inf,Inf,'AbsTol',0.0001);
```

Here, the first line defines `Sintegrand` as the product of `psi1` and `psi2` as a function of x, y, and z. Then, the second line carries out the actual integral, with "`-Inf,Inf,-Inf,Inf,-Inf,Inf`" giving the limits of the integral along x, y, and z as being $-\infty$ to ∞. The "`'AbsTol',0.0001`"

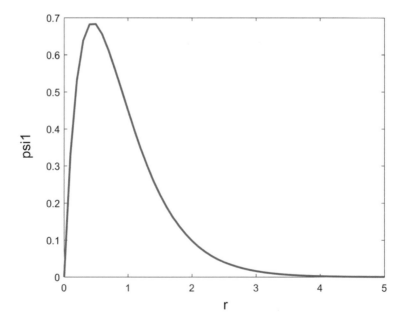

FIGURE 3.2 Plot of `psi1` from $x = 0$ to 5 a_o (see text).

option instructs the function to consider the integral to be converged when it determines the value to an accuracy of 0.0001 (which will be more than sufficient for our purposes). These steps are combined in the following function, STO_Sij:

```
function S_ij = STO_Sij(atom_pos1,coeff1,...
    STO_param1,atom_pos2,coeff2,STO_param2)

psi1 = @(x,y,z) STO_psi(x,y,z,atom_pos1,...
        STO_param1, coeff1);
psi2 = @(x,y,z) STO_psi(x,y,z,atom_pos2,...
        STO_param2, coeff2);

Sintegrand = @(x,y,z) psi1(x,y,z).*psi2(x,y,z);
S_ij = integral3(Sintegrand,-Inf,Inf,-Inf,Inf,...
        -Inf,Inf,'AbsTol',0.0001);
```

Exercise 3.2. Annotate your version of the code above to explain each of the input parameters and the purpose of each line within the program.

EXERCISES FOR EXPLORING OVERLAP INTEGRALS

Now that we have completed the necessary programming for calculating overlap integrals, we are ready to explore how orbitals overlap with each other. In general the overlap integral will depend on two parameters: the distance between the two orbitals and their relative orientations. We'll look at distance first.

Consider two H atoms approaching each other at a distance d along the x-axis. How does the overlap between the s orbitals of these atoms change as a function of d? To calculate this, we can loop over a grid of values for d and calculate a corresponding grid of overlaps:

```
>> H_param = [1 -7.528 1.4592 0 0.0 0.0];
>> x_j = 0;
>> for d = 0:0.2:10
x_j=x_j+1;
S_ij(x_j) = STO_Sij([0,0,0],[1 0 0 0],H_param,...
        [d,0,0],[1 0 0 0],H_param);
end
```

Don't be surprised if this loop takes some time. Numerical triple integrals over an infinite space can be time-consuming. You can speed up the code by replacing the "Inf" numbers with finite ones that are sufficiently large to span a range that captures most of the electron density in the orbitals. Another approach to expediting this calculation would be to coarsen the grid of d values.

Once the loop over d has completed, we plot the $S_{ij}(d)$ function as follows:

```
>> figure
>> plot(0:0.2:10,S_ij,'linewidth',2)
>> set(gca,'FontSize',13)
>> title('\fontsize{20} H 1s \cdot\cdot\cdot H 1s')
>> xlabel('\fontsize{20} d (a_o)')
>> ylabel('\fontsize{20} S_{ij}')
```

where we have added some commands to provide better annotation for the plot. The "set(gca,'FontSize',13)" command changes the font size for the numerical labels along the axes, while the title, xlabel, and ylabel commands provide text for the title and axis labels on the graph. The strings "\fontsize{20}" and "\cdot" make use of MATLAB's ability to interpret some LaTeX commands (the help documentation provides an impressive list of special characters and type-setting options that can be accessed in this way).

The plot resulting from these commands is shown in Figure 3.3. Here we see that the overlap is highest when the distance between the H atoms is zero, where $S_{ij} = 1$. This observation, of course, is nothing more than confirmation that the H 1s orbital is normalized. As the distance is increased, the overlap quickly dies off. At 1.0 a_o (= 0.529 Å), the overlap is already down to 0.74, while at 2.0 a_o the overlap has dropped to 0.37. By 6.0 a_o, the overlap is only 0.006. One should qualify these numbers by recalling that these overlaps are for model H atoms parameterized for a simple Hückel model of benzene. Different numbers, but the same qualitative behavior, would be found for the true H atom wavefunctions we discussed in the last chapter.

Exercise 3.3. Calculate the analogous curves for the in-phase interactions between: (a) a C 2s orbital with a C 2s orbital, (b) a C $2p_x$ orbital with a C $2p_x$ orbital, (c) a C 2s orbital with a C $2p_x$ orbital, and (d) a C $2p_y$ orbital with a C $2p_y$ orbital. To help visualize the interactions being plotted, use the function isosurface_STO_orbital to plot the two orbitals

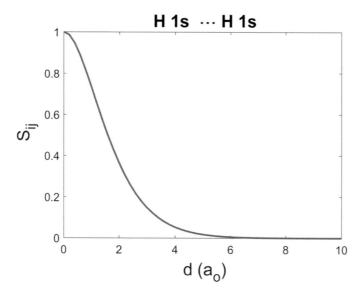

FIGURE 3.3   The overlap integral between two H 1s orbitals as a function of the internuclear distance, d.

involved for each pair at a distance of 3 a_o. An example of a sequence of commands to make such an orbital plot is given here:

```
isosurf = 0.25;
figure
% Place atoms at x = +/- 1.5 to stay within xyz
% grid used by function
isosurface_STO_orbital([-1.5 0 0], C_param,...
    [1 0 0 0], isosurf);
isosurface_STO_orbital([ 1.5 0 0], C_param,...
    [1 0 0 0], isosurf);
% Draw dotted line to show interatomic spacing.
plot3([-1.5 1.5],[0 0],[0 0],':','linewidth',...
    2,'color',[0 0 0]);
```

Comment on the relative magnitudes and distance dependences across the series of overlap integral functions.

The results that you should obtain in Exercise 3.3 are plotted in Figure 3.4. Some trends to note here are that whereas the π-overlap of the p_y functions goes smoothly from 1 to 0 as d increases, the σ-overlap of the p_x functions has a more complicated shape. Because the atomic orbitals have nodes perpendicular to the interatomic axis, the sign of the overlap

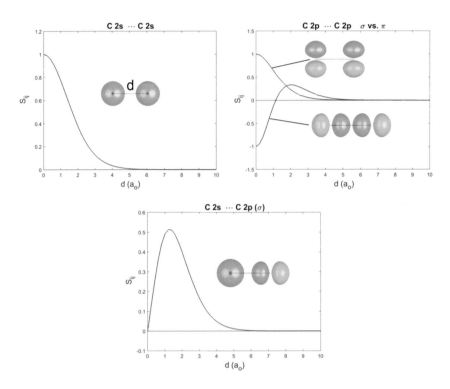

FIGURE 3.4   Overlap integrals between various Slater-type orbitals as functions of internuclear distance.

depends on the relative positions of the atoms. To have in-phase overlap at long-distances means that at extremely short distances (too short to be physically achievable), the functions go out of phase. At $d = 0$, the p_x-p_x overlap is simply -1. Another noteworthy feature is that the maximum for σ overlap is actually higher for a 2s-2p interaction than for a 2p-2p interaction.

Exercise 3.4. Now let's look at the angular dependence of orbital overlap. Consider two carbon atoms: Atom 1 is at the origin, while atom 2 is at a distance of 3 a_o from the origin in the xy-plane at an angle ϕ away from the x-axis. Plot the dependence of S_{ij} on the angle ϕ between a $2p_x$ orbital on atom 1 and a 2s orbital on atom 2, using the range $0° \leq \phi \leq 180°$. You should obtain a plot similar to the one shown in Figure 3.5. In your answer include an annotated version of your code for making this plot. Also explain the shape of the curve, including why the function goes to 0 at $\phi = 90°$.

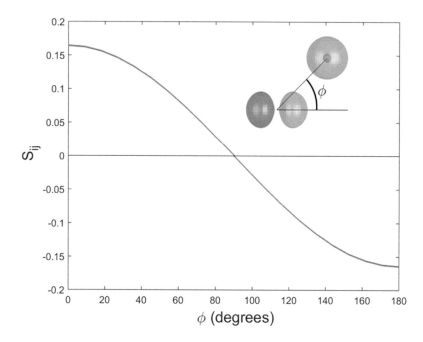

FIGURE 3.5   Angular dependence of the overlap between an s and p orbital.

REFERENCES

1. Yannello, V. J.; Kilduff, B. J.; Fredrickson, D. C. Isolobal Analogies in Intermetallics: The Reversed Approximation MO Approach and Applications to CrGa$_4$- and Ir$_3$Ge$_7$-Type Phases, *Inorg. Chem.* **2014**, *53*, 2730–2741.
2. Stacey, T. E.; Fredrickson, D. C. Perceiving Molecular Themes in the Structures and Bonding of Intermetallic Phases: The Role of Hückel Theory in an *ab initio* Era, *Dalton Trans.* **2012**, *41*, 7801–7813.

CHAPTER 4

Introduction to Molecular Orbital Theory

INTRODUCTION

Through our programming to this point, we have used MATLAB® to visualize the radial and angular components of atomic orbitals, as well as the overlaps between the atomic orbitals of neighboring atoms. Now that we've built up some familiarity with their shapes, and constructed some MATLAB code for graphing them, we can start to look at how interactions between these orbitals build up the electronic structure of molecules. For this quantum mechanical analysis, we will be making some approximations that are common to the simple and extended Hückel methods. First, we assume that the electrons are going to occupy individual orbitals, which are occupied in the ground state through an AUFBAU process from the lowest to the highest energies, analogous to that we use for atoms. Electron-electron interactions, which lead to correlations between the electron positions (and the inability to consider the electrons as independent from each other), will be considered an effect we can add later through perturbation theory if we desire to.

The second simplification that we will take is to consider that these molecular orbitals ψ are built from combining atomic orbitals (ϕ_j's) centered on the molecule's nuclei:

$$\psi = \sum_{j}^{\text{all AOs}} c_j \phi_j$$

This equation is known as the **Linear Combination of Atomic Orbitals (LCAO)** approximation, and could be used to a high degree of accuracy if a sufficiently large number of atomic orbitals (including high energy Rydberg states) are included in the basis set (the palette of functions that is used in constructing the wavefunctions of a system). As we are more interested in a qualitative understanding of electronic structure, we'll use **minimal basis sets**: The smallest set of valence atomic orbitals needed to capture the essence of the bonding. For example, in analyzing an H_2 molecule, our basis set will consist of a H 1s orbital centered on each H nucleus.

Once we have established our basis set, the problem at hand is to determine the best set of coefficients for approximating the true wavefunction of the system. Our guide to determining these coefficients is the **variational principle**, which simply states that any approximation to the ground state of a system will necessarily have a higher expectation value for the energy, <E>, than the ground state itself. The lower the <E> value is for any guess at the ground state wavefunction, the closer that guess is to the true answer. For any given basis set then, the best approximations to the true wavefunctions will be obtained when the coefficients are chosen so that $\dfrac{\partial \langle E \rangle}{\partial c_j} = 0$ for all c_j.

Through the evaluation of these derivatives, we can obtain a general recipe for obtaining these coefficients that can be implemented in MATLAB. We start out with the expression for <E> in terms of the c_j coefficients:

$$\langle E \rangle = \frac{\langle \psi | \hat{H} | \psi \rangle}{\langle \psi | \psi \rangle} = \frac{\left(\sum_i c_i^* \langle \phi_i | \right) \hat{H} \left(\sum_j c_j | \phi_j \rangle \right)}{\left(\sum_i c_i^* \langle \phi_i | \right) \left(\sum_j c_j | \phi_j \rangle \right)}$$

$$= \frac{\sum_i \sum_j c_i^* c_j \langle \phi_i | \hat{H} | \phi_j \rangle}{\sum_i \sum_j c_i^* c_j \langle \phi_i | \phi_j \rangle} = \frac{\sum_i \sum_j c_i^* c_j H_{ij}}{\sum_i \sum_j c_i^* c_j S_{ij}}$$

where $H_{ij} = \langle \phi_i | \hat{H} | \phi_j \rangle$ and $S_{ij} = \langle \phi_i | \phi_j \rangle$ are known as Hamiltonian and Overlap matrix elements, respectively. The denominator on the right-hand side of the equation is there to ensure that the wavefunction stays normalized as the coefficients are varied in search of the minimum <E>. Next, we multiply both sides by this denominator to make the upcoming derivatives simpler to evaluate:

$$\langle E \rangle \sum_i \sum_j c_i^* c_j S_{ij} = \sum_i \sum_j c_i^* c_j H_{ij}$$

Then we can take the derivative of both sides with respect to one of the coefficients c_k^*:

$$\left(\frac{\partial \langle E \rangle}{\partial c_k^*} \right) \sum_i \sum_j c_i^* c_j S_{ij} + \langle E \rangle \frac{\partial \left(\sum_i \sum_j c_i^* c_j S_{ij} \right)}{\partial c_k^*} = \frac{\partial \left(\sum_i \sum_j c_i^* c_j H_{ij} \right)}{\partial c_k^*}$$

Since we're aiming for the minimum <E>, the first term on the left-hand side is simply zero, and we obtain:

$$\langle E \rangle \sum_j c_j S_{kj} = \sum_j c_j H_{kj}$$

which can be written in the form:

$$\sum_j c_j H_{kj} - \langle E \rangle \sum_j c_j S_{kj} = \sum_j c_j \left(H_{kj} - \langle E \rangle S_{kj} \right) = 0$$

If we have N basis functions, we will end up with N of these equations by taking derivatives with respect to each of the N c_k^* coefficients.

There is a standard way to solve such a system of N coupled linear equations. It starts with constructing the **secular determinant**, the determinant of the matrix obtained by using each of the factors $(H_{kj} - \langle E \rangle S_{kj})$ as matrix elements, and setting it to 0:

$$\begin{vmatrix} H_{11} - \langle E \rangle & H_{12} - \langle E \rangle S_{12} & H_{13} - \langle E \rangle S_{13} & \cdots & H_{1n} - \langle E \rangle S_{1n} \\ H_{21} - \langle E \rangle S_{21} & H_{22} - \langle E \rangle & H_{23} - \langle E \rangle S_{23} & \cdots & H_{2n} - \langle E \rangle S_{2n} \\ H_{31} - \langle E \rangle S_{31} & H_{32} - \langle E \rangle S_{32} & H_{33} - \langle E \rangle & \cdots & H_{3n} - \langle E \rangle S_{3n} \\ \vdots & \vdots & \vdots & \ddots & \vdots \\ H_{n1} - \langle E \rangle S_{n1} & \cdots & \cdots & \cdots & H_{nn} - \langle E \rangle \end{vmatrix} = 0$$

or more simply as: $\left| \mathbf{H} - \langle E \rangle \mathbf{S} \right| = 0$, where $\mathbf{H}$ and $\mathbf{S}$ are the Hamiltonian and Overlap matrices. After evaluating the determinant, one solves for $\langle E \rangle$, and then uses $\langle E \rangle$ to go back to the coupled linear equations to solve for the c_j values. This process is equivalent to solving for the eigenvectors of the equation $\mathbf{Hc} = \langle E \rangle \mathbf{Sc}$, which can be easily done with MATLAB once the matrices are constructed.

Obtaining the molecular orbitals through these equations with a minimal basis set (with the Hamiltonian matrix elements determined as described below) is known as the **extended Hückel** method. A further simplification can be made by considering the atomic orbitals to be an orthonormal set so that the S matrix is simply the identify matrix. The eigenvalue problem then becomes $\mathbf{Hc} = \langle E \rangle \mathbf{c}$ which bears a close resemblance to the Schrödinger equation:

$$\mathbf{H}\psi_m = \begin{pmatrix} H_{11} & H_{12} & H_{13} & \cdots & H_{1n} \\ H_{21} & H_{22} & H_{13} & \cdots & H_{2n} \\ H_{31} & H_{32} & H_{33} & \cdots & H_{3n} \\ \vdots & \vdots & \vdots & \ddots & \vdots \\ H_{n1} & \cdots & \cdots & \cdots & H_{nn} \end{pmatrix} \begin{pmatrix} c_{m,1} \\ c_{m,2} \\ c_{m,3} \\ \vdots \\ c_{m,n} \end{pmatrix} = E_m \begin{pmatrix} c_{m,1} \\ c_{m,2} \\ c_{m,3} \\ \vdots \\ c_{m,n} \end{pmatrix} = E_m \psi_m$$

This stripped down version of the extended Hückel approach is referred to as the simple Hückel method.

In this simplest level of theory, the analysis of a compound can be carried out by constructing the Hamiltonian matrix, determining its eigenvectors, and then interpreting the eigenvectors in terms of their bonding, nonbonding, and/or antibonding character. We lose some aspects of the bonding by oversimplifying the overlap matrix, but this drawback is more than offset by the transparency we achieve by doing so. This chapter and the next will demonstrate these ideas using simple diatomic molecules as examples, with MATLAB providing numerical support.

CONSTRUCTION OF THE HAMILTONIAN MATRIX

The first step in solving for the molecular orbitals (MOs) of a molecule is to set up the Hamiltonian matrix describing the interactions between the atomic orbitals in the basis set. The diagonal elements of the matrix are interpreted as the ionization energy of an electron in the atomic orbitals. These elements are simply populated by the H_{ii} parameters we specify for each Slater-type orbital. The off-diagonal elements H_{ij} (with $i{\neq}j$), on the other hand, represent the interaction strengths between different atomic orbitals in the system. A convenient formula for these matrix elements is given by the Wolfsberg-Helmholz approximation:[1]

$$H_{ij} \approx K \frac{\left(H_{ii} + H_{jj}\right)}{2} S_{ij} \text{ with } K = 1.75 \text{ (traditionally)}$$

This approximation makes qualitative sense in that the interaction strength between two orbitals is expected to be proportional to the overlap S_{ij} between them, and an electron shared between these orbitals would be expected to have an ionization energy similar to the average of the two atomic orbital energies $\left(\dfrac{H_{ii} + H_{jj}}{2}\right)$.

The Wolfsberg-Helmholtz approximation is also useful in that we can implement it in a new function based on the STO_Sij code we developed in the last problem set:

```
function H_ij = STO_Hij(atom_pos1,coeff1,...
    STO_params1,atom_pos2,coeff2,STO_params2)

atom_pos1 = atom_pos1/0.5291772;
% convert Angstroms to Bohr.

psi1 = @(x,y,z) STO_psi(x,y,z,atom_pos1,...
STO_params1, coeff1);
psi2 = @(x,y,z) STO_psi(x,y,z,atom_pos2,...
STO_params2, coeff2);

Sintegrand = @(x,y,z) psi1(x,y,z).*psi2(x,y,z);
S_ij = integral3(Sintegrand,-Inf,Inf,-Inf,Inf,...
-Inf,Inf,'AbsTol',0.0001);

K = 1.75; % Wolfsberg-Helmholz constant.

% STO_parameters = [n_s H_ii(s) zeta_s n_p H_ii(p)
%zeta_p]
H_ii_s = STO_params1(1,2);
H_ii_p = STO_params1(1,5);
H_ii = norm(coeff1(1))^2*H_ii_s;
```

```
H_ii = H_ii + (norm(coeff1(2))^2+norm(coeff1(3))...
^2+norm(coeff1(4))^2)*H_ii_p;
% H_ii is weighted average of H_ii_s and H_ii_p,
% in someone wants to use hybrid orbitals. Be
% careful with the placement of the parentheses in
% front of H_ii_p; this is a common place for bugs
% to sneak in.

H_jj_s = STO_params2(1,2);
H_jj_p = STO_params2(1,5);
H_jj = norm(coeff2(1))^2*H_jj_s;
H_jj = H_jj + (norm(coeff2(2))^2+norm(coeff2(3))...
^2+norm(coeff2(4))^2)*H_jj_p;

H_ij = K*0.5*(H_ii+H_jj)*S_ij;
```

Here, the parts highlighted in black are added to or modified from the STO _ Sij function. In the lines for setting up the H_{ii} and H_{jj} numbers, we have used weighted combinations of the s and p orbital H_{ii} values according to the coeff1 and coeff2 vectors. With these formulas in place, we open the possibility of calculating not only Hamiltonian matrix elements between atomic orbitals but also between hybrid orbitals. Note also that a conversion factor from Ångstroms to Bohr has been added for the atom _ pos1 and atom _ pos2 variables. This change allows us to express molecular geometries in Ångstroms with the actual calculations using Bohr. *Add this unit conversion code to your STO_Sij function as well.*

Now that we have this function for calculating matrix elements, we are in a position to set up the full Hamiltonian matrix for the H_2 molecule. Let's consider the H atoms to be 0.7 Å apart, with atom 1 at (*x*, *y*, *z*) = (0,0,0) and atom 2 at (0.7 Å, 0, 0). We'll start with an extended Hückel treatment, in which the standard H atom parameters can be entered with:

```
>> H_param_eH = [1 -13.600 1.300 0 0.0 0.0];
```

The Hamiltonian matrix elements between the 1s orbitals on atom 1 and atom 2 can then be built up as a 2 × 2 matrix.

```
>> H_eH(1,1) = -13.6;
>> H_eH(2,2) = -13.6;
>> s = [1 0 0 0]; % Abbreviation for coefficients
                  % for s orbital
>> H_eH(1,2) = STO_Hij([0 0 0],s,H_param_eH,...
                [0.7 0 0],s,H_param_eH);
>> H_eH(2,1) = STO_Hij([0 0 0],s,H_param_eH,...
                [0.7 0 0],s,H_param_eH);
>> H_eH

H_eH =

  -13.6000 -15.7971
  -15.7971 -13.6000
```

The overlap matrix is constructed in a similar fashion:

```
>> S_eH(1,1) = STO_Sij([0 0 0],s,H_param_eH,...
    [0  0 0],s,H_param_eH);
>> S_eH(1,2) = STO_Sij([0 0 0],s,H_param_eH,...
    [0.7 0 0],s,H_param_eH);
>> S_eH(2,1) = STO_Sij([0 0 0],s,H_param_eH,...
    [0.7 0 0],s,H_param_eH);
>> S_eH(1,1) = STO_Sij([0.7 0 0],s,H_param_eH,...
    [0.7 0 0],s,H_param_eH);
>> S_eH(2,2) = STO_Sij([0.7 0 0],s,H_param_eH,...
    [0.7 0 0],s,H_param_eH);
>> S_eH

S_eH =

  1.0000  0.6637
  0.6637  1.0000
```

SOLVING FOR THE MOLECULAR ORBITALS

Now that we have set-up the Hamiltonian (H) and Overlap (S) matrices for H_2, the MOs are obtained by solving the eigenvalue problem:

$$Hc = ESc$$

where c is one of the desired eigenvectors whose coefficients represent the contributions of the atomic orbitals to the molecular orbitals. Problems of this form are solvable in MATLAB using the eig function. For example, the command eig(H_eH, S_eH) returns the possible eigenvalues E for the system:

```
>> eig(H_eH, S_eH)

ans =

-17.6692
6.5338
```

This output tells us that there are two eigenvalues for E, one at −17.67 eV (stabilized by about 4 eV relative to the original H 1s orbital energy) and the other at +6.53 (destabilized by 20 eV relative to the atomic orbital energy).

The eig function will also give the eigenvectors corresponding to these energies, if we ask for them:

```
>> [psi_H2, E_H2] = eig(H_eH, S_eH)

psi_H2 =

-0.5482   1.2194
-0.5482  -1.2194


E_H2 =

-17.6692      0
      0  6.5338
```

Here, the matrix psi_H2 contains the eigenvectors as column vectors. The atomic orbital coefficients for the first MO can be found by typing:

```
>> psi_H2(:,1) % ":" means all elements along the
   % given row or column.

ans =
```

```
    -0.5482
    -0.5482
```

This corresponds to an in-phase combination of the H 1s orbitals. Likewise, the coefficients of the second MO are obtained as:

```
>> psi_H2(:,2)

ans =

    1.2194
   -1.2194
```

This is simply the out-of-phase combination of the H 1s orbitals.

In addition to the eigenvectors, the MATLAB function has also returned E _ H2. This is the Hamiltonian matrix written in the basis of these eigenvectors, and as such its diagonal elements are the eigenvalues, and its off-diagonal elements are all zero.

Exercise 4.1. Show that calculating the Hamiltonian matrix elements between the eigenvectors leads to the matrix E _ H2. For example, the interaction between MO1 and MO1 can be calculated through the matrix equivalent of the equation $H(\text{MO basis})_{11} = \langle \text{MO1} | \hat{H} | \text{MO1} \rangle$. In MATLAB, this could take the form:

```
>> H_MObasis(1,1) = psi_H2(:,1)'*H_eH*psi_H2(:,1)
          % ' = matrix transpose

H_MObasis =

   -17.6692
```

Calculate the full H_MObasis matrix, and show that it is equivalent to E _ H2 above.

VISUALIZING ISOSURFACES OF MOS IN MATLAB

How might we draw a picture of the MOs we've obtained in the previous section? One way is to draw isosurfaces of the molecular orbitals. Doing this requires only simple modifications to the isosurface_STO_orbital function we wrote for the last chapter:

```
function isosurface_MO(atom_pos,STO_params,...
    coeff, psi_value)
% In atom_pos, STO_params, and coefficients values
% for the atoms of the molecule are stacked as rows.
atom_pos = atom_pos/0.5291772;
% convert Angstroms to Bohr.

% Set up grid in Cartesian space
x_j = 0;
y_j = 0;
z_j = 0;
for x = -4:.1:4
  x_j = x_j+1;
  y_j=0;
  for y = -4:.1:4
    y_j = y_j+1;
    z_j = 0;
    for z = -4:.1:4
      z_j = z_j+1;
      X_grid(x_j,y_j,z_j) = x;
      Y_grid(x_j,y_j,z_j) = y;
      Z_grid(x_j,y_j,z_j) = z;
    end
  end
end

num_atoms = size(atom_pos);
num_atoms = num_atoms(1);

grid_size = size(X_grid);
psi_grid = zeros(grid_size);

for j = 1:num_atoms
  % Add up contributions from different atoms.
  psi_grid=psi_grid+STO_psi(X_grid,Y_grid,Z_grid,...
    atom_pos(j,:),STO_params(j,:),coeff(j,:));
end

for iso_level = psi_value
  psi_surface = isosurface(X_grid,Y_grid, Z_grid,...
        psi_grid, iso_level);
```

```
    patch(psi_surface,'FaceColor',[0 90 255]/255,...
            'FaceAlpha',.2,'EdgeColor','none');
% FaceAlpha sets transparency of surface:
% 1 = solid, 0 = invisible.
end

hold on

% Draw negative isosurfaces
for iso_level = psi_value
   psi_surface = isosurface(X_grid,Y_grid, Z_grid,...
            psi_grid, -iso_level);
   patch(psi_surface,'FaceColor',[0 140/255 0],...
            'FaceAlpha',.2,'EdgeColor','none');
end

axis equal  % set scales along x, y, and z to be
            % equal
axis off
```

The major change here is that the input variables atom_pos, STO_params, and coeff, can now all have more than one row, with each row corresponding to a different atom. The wavefunction is then built up by looping over these rows and adding up the atomic orbitals on a grid. Here, we have also added the unit conversion at the beginning so that the molecular geometry can be specified in Ångstroms (the units in the plot will still be in Bohr, but we can't see this when the axes are not drawn). In addition, we have added loops around the generation of the isosurfaces so that psi_value can be a row of isosurface values to be plotted together.

We're almost ready now to plot the isosurfaces of the MOs. The next step is to set up the input variables for the isosurface_MO function. This involves rearranging some information we already have:

```
>> p_components = [0 0 0; 0 0 0];
>> coeff = [psi_H2(:,1) p_components]
% need to add p orbital coefficients

coeff =

   -0.5482    0     0     0
   -0.5482    0     0     0
```

```
>> atom_pos = [ 0 0 0 ; 0.7 0 0 ];
>> STO_params = [H_param_eH; H_param_eH]

STO_params =

   1.0000 -13.6000  1.3000      0      0      0
   1.0000 -13.6000  1.3000      0      0      0
```

Then a series of isosurfaces as shown can be generated (see Figure 4.1).

```
>> figure
>> isosurface_MO(atom_pos,STO_params,coeff,...
   0.05:0.05:0.3)
>> light
```

In this image, we can clearly see that the in-phase combination of 1s orbitals has led to constructive interference in the space between the two nuclei. This function is thus considered to be a bonding MO.

One issue with the picture that we have just made is that the positions of the nuclei are very difficult to make out. We can solve this by plotting little spheres on the nuclear positions, pulling some of our code from the draw_atomic_orbital_xyz.m function we made in Chapter 2:

```
function draw_molecule_spheres(atom_pos,color,...
   sphere_radius,sphere_resolution)
atom_pos = atom_pos/0.5291772;
% convert Angstroms to Bohr.
sphere_radius = sphere_radius/0.5291772;
```

FIGURE 4.1   The bonding MO of H_2 calculated with the extended Hückel method and visualized through concentric isosurfaces.

```
num_atoms = size(atom_pos);
num_atoms = num_atoms(1);
for j = 1:num_atoms
 [x,y,z] = sphere(sphere_resolution);
 [phi, theta_matlab, r] = cart2sph(x,y,z);
 theta = pi/2 - theta_matlab;

 r_new = sphere_radius;

 [x_new, y_new, z_new] = sph2cart...
                       (phi,theta_matlab,r_new);
 x_new = x_new + atom_pos(j,1);
 y_new = y_new + atom_pos(j,2);
 z_new = z_new + atom_pos(j,3);
 surf(x_new,y_new,z_new,'FaceColor',color,...
                     'EdgeColor','none');
end

axis equal;
```

We can then use the function as follows to overlay the isosurfaces onto a picture of the molecule:

```
>> figure
>> isosurface_MO(atom_pos,STO_params,coeff,...
   0.05:0.05:0.3)
>> draw_molecule_spheres(atom_pos,[0,0,0],.2,40)
>> light
```

These commands should then yield the image in Figure 4.2.

FIGURE 4.2 The bonding MO of H_2 with spheres included to mark the nuclear positions.

Exercise 4.2. Reproduce the picture in Figure 4.3 showing the higher energy MO. Explain why it is higher in energy than the bonding MO, and why this is referred to as an antibonding MO.

Exercise 4.3. Use the energies of the atomic and molecular orbitals to create a MO diagram such as that shown in Figure 4.4. Note that you don't need to use MATLAB to generate the whole picture in one step, but you

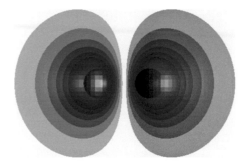

FIGURE 4.3   The antibonding MO of H_2.

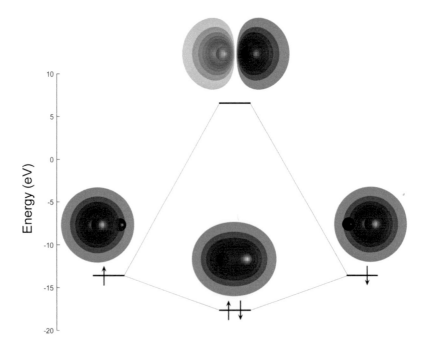

FIGURE 4.4   MO diagram for the formation of an H_2 molecule from two H atoms.

are free to combine and add the various elements using programs such as PowerPoint or Photoshop.

In this diagram, we also include the electrons brought to the molecule by the H atoms, one electron for each hydrogen atom. In the MOs, the electrons fill in the energy levels from lowest to highest so that the lower bonding level is filled and the upper antibonding level is empty. At this simple level of theory, in which electron-electron interactions are ignored, the total energy of the molecule is written as the sum of the electron energies. In this case, the energy would be -17.67 eV $\times$ 2.

A comparison of the MO energies in the above diagram with the original AO levels makes it very clear that the bonding level has been stabilized by a much smaller amount than the antibonding level is destabilized. This effect is a result of the overlap between the atomic orbitals and the normalization of the wavefunctions:

$$\langle \psi \mid \psi \rangle = \sum_i \sum_j c_i^* c_j \langle \phi_i \mid \phi_j \rangle = \sum_i |c_i|^2 + \sum_i \sum_{j \neq i} c_i^* c_j S_{ij} = 1$$

When the $c_i^* c_j S_{ij}$'s are positive, as in a bonding function, the coefficients must be smaller (relative to the case of an orthogonal basis set) to keep the function normalized. These smaller coefficients dampen the amount of bonding present in the function, which for each interaction is $c_i^* c_j H_{ij} + c_j^* c_i H_{ji}$. In the antibonding case, on the other hand, the $c_i^* c_j S_{ij}$'s are negative. The coefficients then need to be larger to make the sum come out to one. These larger coefficients in turn exacerbate the effect of the antibonding interactions on the MO energy.

The asymmetry between energy changes of bonding and antibonding interactions has an important consequence: Interactions between two filled levels are net repulsive in nature. Included here would be the interactions between two C-H bonds, or two He atoms, systems in which we know steric repulsion takes over at short distances.

EXTENDED HÜCKEL VS. SIMPLE HÜCKEL

So far, we have carried out all of our analysis using the extended Hückel method, in which the MOs are the solutions to the eigenvalue problem $\mathbf{Hc} = \mathbf{ESc}$. How does the answer change when we move to the simple Hückel method, where we solve $\mathbf{Hc} = \mathbf{Ec}$ instead, as though the atomic orbitals comprise an orthonormal basis set? To answer this question, we can simply use the eig function without including the overlap matrix:

```
>> [psi_H2, E_H2] = eig(H_eH)

psi_H2 =

    0.7071   -0.7071
    0.7071    0.7071

E_H2 =

   -29.3971        0
         0    2.1971
```

In the psi _ H2 output, we still have a bonding function based on a symmetric sum of the two 1s orbitals, and an antibonding function based on their difference. However, rather than having small coefficients for the bonding function and larger ones for the antibonding function, all of the coefficients are $\pm \dfrac{1}{\sqrt{2}}$ so that the functions are normalized with $c_1^2 + c_2^2 = 1$.

Differences from the extended Hückel result are also apparent in the MO energies. The Hückel bonding orbital is more than 11 eV lower in energy than its extended Hückel counterpart, while the antibonding function is more than 4 eV lower than that in the extended Hückel calculation. This trend is fairly general: Removing the overlap matrix tends to over-emphasize bonding and deemphasize antibonding relative to the extended Hückel results.

Another difference is in how these MO energies compare with the original 1s orbital energies, as plotted in Figure 4.5. A quick comparison between the MO energies and those of the original atomic orbitals shows that the stabilization bonding MO and destabilization of the antibonding level is simply $|H_{ij}| = 15.7971$ eV. In other words, at the Hückel level the MO energies are just $H_{ii} \pm H_{ij}$. The bonding and antibonding levels are shifted from the free atomic orbital energies by equal magnitudes during the interaction, which explains the large differences in the energies from the extended Hückel results.

The simple Hückel method can give realistic results, however, if we adjust the parameterization. Consider the H parameters we used before that were refined to reproduce the energy levels of crystalline benzene correctly:

```
>> H_param_sH = [1 -7.528 1.4592 0 0.0 0.0];
```

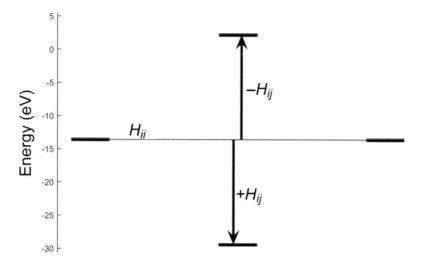

FIGURE 4.5 Energy levels for the H_2 MO diagram calculated using the simple Hückel method.

We can set up a Hamiltonian matrix using these parameters as follows:

```
>> H_sH(1,1)  =  -7.528;
>> H_sH(2,2)  =  -7.528;
>> H_sH(1,2)  =  STO_Hij([0 0 0],s,H_param_sH,...
   [0.7 0 0],s,H_param_sH);
>> H_sH(2,1)  =  H_sH(1,2);
>> H_sH

H_sH =

   -7.5280   -7.9760
   -7.9760   -7.5280
```

And then recalculate the MO energies:

```
>> eig(H_sH)

ans =

  -15.5040
    0.4480
```

The result is a bonding MO that is much more similar in energy to that of the extended Hückel calculation (and also more realistic). For the anti-bonding levels, the energies are still quite different from each other, but in general this type of theoretical calculations is not expected to reproduce the energies of unoccupied levels very accurately.

Some insights can be gleaned by comparing the extended and simple Hückel parameters. Note that the H_{ii} value is significantly less negative and the ζ value is larger in the simple Hückel parameters. These changes both serve to decrease the magnitudes of the H_{ij} values (see the Wolfsberg-Helmholz formula) so that the bonding MO is not over-estimated. One result of this is that the H_{ii} values no longer represent realistic atomic orbital ionization energies. Instead they are shifted upwards, which can be interpreted as an energetic penalty for orthogonalizing the atomic orbital to the atomic orbitals on neighboring atoms.

Despite these differences, working with the simple Hückel method can bring similar results to those of the extended Hückel method with much simpler mathematics. We will thus use the simple Hückel method for the rest of this book, while being aware of its limitations, such as the one illustrated in the next exercise.

Exercise 4.4. *A big problem with Hückel theory.* What do we lose upon going from extended Hückel to simple Hückel? Calculate the simple Hückel MO energies for a series of H_2 and H_2^{2-} molecules with the bond distances ranging from 0.1 Å to 2.0 Å. Plot the total energy of the molecule ($E_{total} = \sum_{j}^{\text{all MOs}} o_j E_j$, were o_j is the occupancy of MO j, which will be 0 or 2 in this case) as a function of bond distance. What is unexpected about the result you obtain? Where is this problem coming from? As you will see from results for the H_2^{2-} problem, Hückel theory is unable to reproduce the net repulsion arising in the interaction between filled levels.[2] Such interactions instead appear as energy neutral. In our interpretation of the results of our calculations, we will then need to add such effects qualitatively on our own.

A SIMPLIFIED REPRESENTATION OF MOs IN MATLAB

Using isosurfaces to represent MOs can become a little cumbersome for larger molecules. Is there a simpler way to plot them? In the last chapter, we built up a function for plotting atomic orbitals centered at any point in space, draw_atomic_orbital_xyz.m, in which the input variables for the function are the atomic position in Cartesian coordinates and the orbital coefficients, e.g.:

```
function draw_atomic_orbital_xyz(x,y,z,s,px,py,...
    pz,dz2,dx2y2,dxy,dxz,dyz);
% x,y,z = Cartesian coordinates of atom.
% s,px,...,dyz = coefficients of corresponding
% orbital in wavefunction.
% Check that the order the the input variables
% here matches that in your
% own version of the function.
```

We can use this function to plot the full MO atom by atom in a simplified representation. Since we are going to be plotting several orbitals in the same window, it is a good idea to take out the "light" command in draw_atomic_orbital_xyz, as each time it is called additional light is added to the plot. Adding the Ångstrom to Bohr conversion for the coordinates would also be a good idea to maintain compatibility between all of your functions.

```
>> figure
>> scaleorb=0.75;
%   set scale of orbital features for plotting
>> draw_atomic_orbital_xyz(0,0,0,scaleorb*0.7071,...
    0,0,0,0,0,0,0,0);
>> hold on
>> draw_atomic_orbital_xyz(0.7,0,0,scaleorb*...
    0.7071,0,0,0,0,0,0,0,0);
>> axis off
>> light
```

You should see an image similar to that displayed in Figure 4.6.

As you can see from the operations above, generating pictures of MOs atom by atom in this way could become tedious. It is then valuable to make a MATLAB function to automate this process. This function would receive as input the Cartesian coordinates of the atoms, the number of atomic orbitals each atom has, the orbital coefficients, and a scale factor to adjust the size of the orbital features. The header for the function could then be:

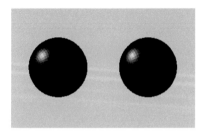

FIGURE 4.6   Bonding MO of H_2 plotted with atomic orbitals scaled according to their coefficients.

```
function drawMO(coordinates, orbital_counts,...
orbital_coefficients, scale)
% coordinates = array with three columns, one each
% for the x,y, and z coordinates
% orbital_counts = array with number of orbitals
% for each atom as a row vector
% orbital_coefficients = column vector containing
% orbital coefficients to plot
% scale = number to multiply orbital_coefficients
% by to adjust scale of the plot
```

Within the function, one then loops over the atoms and plots the orbital on each of them:

```
figure % start new figure
natoms=size(coordinates);
natoms=natoms(1);
orbitals_used=0;
for j = 1:natoms
  s=0; px=0; py=0; pz=0; dx2=0; dz2=0; dxy=0;...
    dxz=0; dyz=0;
  if orbital_counts(j)==1  % s orbital only
    s=scale_factor*orbital_coefficients...
      (orbitals_used+1);
    orbitals_used=orbitals_used+1;
  end
  if orbital_counts(j)==4 % s+p orbitals
    s=scale_factor*orbital_coefficients...
      (orbitals_used+1,1);
```

```
      px=scale_factor*orbital_coefficients...
          (orbitals_used+2,1);
      py=scale_factor*orbital_coefficients...
          (orbitals_used+3,1);
      pz=scale_factor*orbital_coefficients...
          (orbitals_used+4,1);
      orbitals_used=orbitals_used+4;
   end
   if orbital_counts(j)==9 % s+p+d orbitals
      s=scale_factor*orbital_coefficients...
          (orbitals_used+1,1);
      px=scale_factor*orbital_coefficients...
          (orbitals_used+2,1);
      py=scale_factor*orbital_coefficients...
          (orbitals_used+3,1);
      pz=scale_factor*orbital_coefficients...
          (orbitals_used+4,1);
      dx2=scale_factor*orbital_coefficients...
          (orbtals_used+5,1);
      dz2=scale_factor*orbital_coefficients...
          (orbtals_used+6,1);
      dxy=scale_factor*orbital_coefficients...
          (orbitals_used+7,1);
      dxz=scale_factor*orbital_coefficients...
          (orbitals_used+8,1);
      dyz=scale_factor*orbital_coefficients...
          (orbitals_used+9,1);
      orbitals_used=orbitals_used+9;
   end
   x=coordinates(j,1); y=coordinates(j,2); ...
   z=coordinates(j,3);
   hold on
   draw_atomic_orbital_xyz(x,y,z,s,px,py,pz,dx2,...
          dz2,dxy,dxz,dyz);
end
```

With this function, the procedure for drawing the orbital pictures is then much simpler. We define arrays for the atomic coordinates, and the number of atomic orbitals for each atom:

```
>> Cart = [0.0 0.0 0.0
   0.7 0.0 0.0];
>> Orbital_list = [1,1];
```

then run drawMO using the orbital coefficients obtained earlier:

```
>> drawMO(Cart, Orbital_list, psi_H2(:,1),0.75);
>> drawMO(Cart, Orbital_list, psi_H2(:,2),0.75);
```

REFERENCES

1. Wolfsberg, M.; Helmholz, L. The Spectra and Electronic Structure of the Tetrahedral Ions MnO4–, CrO4––, and ClO4–, *J. Chem. Phys.* **1952**, *20*, 837–843.
2. Lee, S. Structural Diversity in Solid State Chemistry: A Story of Squares and Triangles, *Annu. Rev. Phys. Chem.* **1996**, *47*, 397–419.

CHAPTER 5

The Molecular Orbitals of N_2

INTRODUCTION

Now that we have seen how bonding works in one of the simplest molecules, let's start working toward more complicated systems. In this chapter, we will consider another homonuclear diatomic molecule, N_2. While N_2 is still very simple, it has some important lessons to teach us about s-p hybridization. Before setting out on this molecule directly, however, there are some preparations we should make: As each of the N atoms brings four valence atomic orbitals to the system, the full set of interactions will create an 8×8 Hamiltonian matrix. Calculating the 64 elements of this matrix one-by-one would not provide much encouragement for working with molecules with more orbitals (which, in fact, includes most molecules). Our first step, then, is to automate the process of constructing the Hamiltonian.

SOLVING THE GENERAL PROBLEM OF BUILDING THE HAMILTONIAN

In the process of calculating the H_{ij} values we also calculate S_{ij}. We can slightly modify the first line of our STO _ Hij function for it to output both of these quantities:

```
function [H_ij, S_ij] = STO_Hij_Sij(atom_pos1,...
coeff1,STO_params1,atom_pos2,coeff2,STO_params2)
```

As we calculate the Hamiltonian matrices for various molecules, we'll then output the overlap matrices as well, since they are essentially free, computationally.

In order to explore a wide variety of molecular systems, we are going to want a generalized function for calculating the Hamiltonian matrix from an input geometry, something of the following form:

```
function [H, S, orb_list] = build_hamiltonian(filename)
```

where `filename` gives the name of a separate file that contains a molecular geometry to be read in by the function. To begin, we'll need to specify a format for the molecular geometry in the input file. Let's use a simple Cartesian x, y, z format as in:

```
N 0.0 0.0 0.0
N 0.0 0.0 1.4
```

in which each line gives the element name and x, y, z coordinates (in Å) for a different atom.

When data in a file is nicely arranged into columns of information such as the list above, it can be easily read by MATLAB® using the `textread` command:

```
[atomname,x,y,z] = textread(filename,'%s %f %f %f');
```

Here `atomname`, `x`, `y`, and `z` are each populated by one of the four columns of the input file, and the string `'%s %f %f %f'` indicates that the first column contains string data (encoded as `%s`, mirroring the syntax of the C programming language) and the next three contain real numbers (encoded as `%f`). At this point we have all of the atomic positions and the element names for each atom read in.

Our next step is to assign Hückel parameters to each of these atoms. To do this, we can compare the element names in `atomname` with those in a list of elements for which we have parameters. Such comparisons are possible in MATLAB with the `strcmp` command. `strcmp(string1, string2)` returns 1 when `string1` = `string2`, and 0 otherwise. Here is an example of such a search:

```
natoms = size(x);
natoms = natoms(1);
```

```
params = zeros(natoms,6);
orb_list = ones(1,natoms);

% Assign STO parameters
num_orbitals = 0;
for j = 1:natoms
  foundit = 0;
  if(strcmp(atomname(j),'H')==1)
      params(j,:) = [1 -7.528 1.4592 0 0.0 0.0];
      num_orbitals = num_orbitals+1;
      orb_list(1,j) = 1;
      foundit = 1;
  end
  if(strcmp(atomname(j),'C')==1)
      params(j,:) = [2 -11.241 2.25 2 -7.857 2.22];
      num_orbitals = num_orbitals+4;
      orb_list(1,j) = 4;
      foundit = 1;
  end
  if(strcmp(atomname(j),'N')==1)
      params(j,:) = [2 -16.507 2.455 2 -9.300 2.397];
      num_orbitals = num_orbitals+4;
      orb_list(1,j) = 4;
      foundit = 1;
  end
  if(foundit == 0)
    fprintf ...
    ('Parameters for atom %d not found.\n',j);
  end
end
```

In this procedure, the variable params consists of a stack for Hückel parameters, with one row for each atom in the structure. As each atom is matched with an element, the number of atomic orbitals it contains gets updated in orb_list, and the total number of atomic orbitals for the molecule gets tallied in num_orbitals. Also for each atom, we use the variable foundit to check whether we have actually assigned parameters to that atom. foundit starts out equal to zero, but gets promoted to one once parameters are assigned. If foundit is still zero after we have tried all of the elements in our parameter list, a warning message will be displayed.

Now that we have the atomic geometry and Hückel parameters in place, all we need to do is calculate the Hamiltonian matrix elements for each pair of atomic orbitals. For this process, we will take a double loop over the j and k in the matrix elements H_{jk} for all atomic orbitals. As we proceed, the tricky part will be keeping track of which orbital belongs to which atom, and which of the atom's possible orbitals it corresponds to. A rather cumbersome code for accomplishing these objectives is shown below:

```
% CALCULATION H AND S MATRICES
H_ao = zeros(num_orbitals, num_orbitals);
S_ao = zeros(num_orbitals, num_orbitals);
atomj = 1;
for j = 1:num_orbitals
  fprintf...
  ('Interactions for orbital%d\n',j);
  % Determine whether we've already gone through
  % all orbitals on atom;
  if j > sum(orb_list(1,1:atomj))
    atomj = atomj + 1;
  end
  % Set position of nucleus orbital sits on.
  atom_pos1 = atom_pos(atomj,:);

  coeff1 = zeros(1,4);
  % Determine which orbital on atom we are dealing
  % with.
  ao_num = j;
  if (atomj > 1)
    ao_num = j - sum(orb_list(1,1:(atomj-1)));
  end
  coeff1(ao_num) = 1;

  atomk = 1;
  for k = 1:num_orbitals
    if k > sum(orb_list(1,1:atomk))
      atomk = atomk + 1;
    end
    % Set position of nucleus orbital sits on.
    atom_pos2 = atom_pos(atomk,:);
    coeff2 = zeros(1,4);
```

```
% Determine which orbital on atom we are
% dealing with.
ao_num = k;
if (atomk > 1)
    ao_num = k - sum(orb_list(1,1:(atomk-1)));
end
coeff2(ao_num) = 1;
if(j==k)
  if(ao_num == 1)
    H_ao(j,j) = params(atomj,2);
    S_ao(j,j) = 1;
  end
  if(ao_num>1)&&(ao_num<5)
    H_ao(j,j) = params(atomj,5);
    S_ao(j,j) = 1;
  end
else
  if(atomj~=atomk)
  % Don't waste time calculating overlaps
  % between orbitals on same atom.
  [H_ao(j,k) S_ao(j,k)] = STO_Hij_Sij...
  (atom_pos1, coeff1,params(atomj,:),...
  atom_pos2,coeff2,params(atomk,:));

  end
 end
end
end
```

Some explanation is certainly needed here: `atomj` and `atomk` keep track of which atom we are on in the `j` and `k` loops, respectively. `sum(orb_list(1,1:atomj))` gives the total number of orbitals for all atoms up through `atomj`, while `sum(orb_list(1,1:(atomj-1)))` gives the total number of orbitals for all of the atoms before `atomj` in the list. `ao_num = j - sum(orb_list(1,1:(atomj-1)))` then yields which of `atomj`'s atomic orbitals `j` corresponds to. You will also notice the special treatment to the diagonal elements of the matrix.

Exercise 5.1. Annotate the above code to explain how it achieves the proper mapping of the `atom_pos1`, `atom_pos2`, `coeff1`, `coeff2`, and `params` variables to the orbital list.

With this last sequence of commands, we now have everything we need for the Hamiltonian generation function. The full function is given here:

```
function [H_ao, S_ao, orb_list] = ...
    build_hamiltonian(filename)

[atomname,x,y,z] = textread...
                    (filename,'%s %f %f %f');

natoms = size(x);
natoms = natoms(1);

atom_pos = [x y z];

params = zeros(natoms,6);
orb_list = ones(1,natoms);

% Assign STO parameters
num_orbitals = 0;
for j = 1:natoms
  foundit = 0;
  if(strcmp(atomname(j),'H')==1)
      params(j,:) = [1 -7.528 1.4592 0 0.0 0.0];
      num_orbitals = num_orbitals+1;
      orb_list(1,j) = 1;
      foundit = 1;
  end
  if(strcmp(atomname(j),'C')==1)
      params(j,:) = [2 -11.241 2.25 2 -7.857 2.22];
      num_orbitals = num_orbitals+4;
      orb_list(1,j) = 4;
      foundit = 1;
  end
  if(strcmp(atomname(j),'N')==1)
      params(j,:) = [2 -16.507 ...
              2.455 2 -9.300 2.397];
      num_orbitals = num_orbitals+4;
      orb_list(1,j) = 4;
      foundit = 1;
  end
  if(foundit == 0)
    fprintf...
    ('Parameters for atom %d not found.\n',j);
```

```
    end
end

% num_orbitals

% CALCULATION H AND S MATRICES
H_ao = zeros(num_orbitals, num_orbitals);
S_ao = zeros(num_orbitals, num_orbitals);
atomj = 1;
for j = 1:num_orbitals
  fprintf ...
          ('Interactions for orbital %d\n',j);
  % Determine whether we've already gone through
  % all orbitals on atom;
  if j > sum(orb_list(1,1:atomj))
    atomj = atomj + 1;
  end
  % Set position of nucleus orbital sits on.
  atom_pos1 = atom_pos(atomj,:);

  coeff1 = zeros(1,4);
  % Determine which orbital on atom we are dealing
  % with.
  ao_num = j;
  if (atomj > 1)
    ao_num = j - sum(orb_list(1,1:(atomj-1)));
  end
  coeff1(ao_num) = 1;

  atomk = 1;
  for k = 1:num_orbitals
    if k > sum(orb_list(1,1:atomk))
      atomk = atomk + 1;
    end
    % Set position of nucleus orbital sits on.
    atom_pos2 = atom_pos(atomk,:);
    coeff2 = zeros(1,4);
    % Determine which orbital on atom we are
    % dealing with.
    ao_num = k;
    if (atomk > 1)
      ao_num = k - sum(orb_list(1,1:(atomk-1)));
    end
```

```
    coeff2(ao_num) = 1;
    if(j==k)
      if(ao_num == 1)
        H_ao(j,j) = params(atomj,2);
        S_ao(j,j) = 1;
      end
      if(ao_num>1)&&(ao_num<5)
        H_ao(j,j) = params(atomj,5);
        S_ao(j,j) = 1;
      end
    else
     if(atomj~=atomk)
  % Don't waste time calculating overlaps between
  % orbitals on same atom.
      [H_ao(j,k) S_ao(j,k)] = STO_Hij_Sij...
         (atom_pos1, coeff1,params(atomj,:),...
         atom_pos2,coeff2,params(atomk,:));
     end
    end
   end
 end
```

THE BRUTE FORCE SOLUTION OF THE MOs OF N_2

We now have the code to calculate the MO diagram of N_2 or any other molecule we may be interested in. In the next two exercises we will test these functions.

Exercise 5.2. **The N_2 Hamiltonian.** Use the build_hamiltonian function to calculate the Hückel Hamiltonian for a N_2 molecule. Orient the molecule with the bond along the z-axis, and set the bond length to 1.4 Å. Take a screenshot of the Hamiltonian and annotate the non-zero elements with a description of the type of interaction they quantify.

Exercise 5.3. **The N_2 MO diagram.** Diagonalize the Hamiltonian matrix to obtain eigenvectors and eigenvalues, then use your functions from the previous chapters to plot pictures of the orbitals. Finally, compile the resulting pictures and energy eigenvalues into an MO diagram into a single image. Include the occupation of the MOs as well. You should obtain a diagram that resembles Figure 5.1.

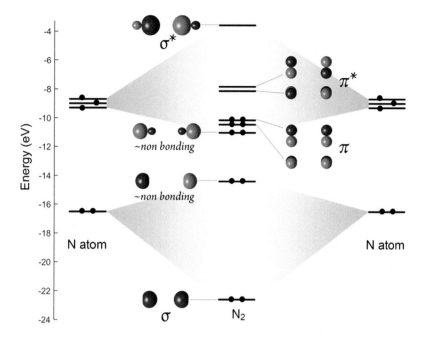

FIGURE 5.1   The MO diagram of N$_2$.

Hint: You can make the generation of the MO pictures much easier by setting the orientation of the molecule with a command (as opposed to rotating it interactively in the Figure window). Try the following commands:

```
view([1,.15,0]);
camup([0,1,0]);
```

which set the direction down which the camera looks, and the vertical direction of the figure, respectively.

In Exercises 5.2 and 5.3, we used a brute-force approach to obtaining the MOs for N$_2$, in which we threw all of the basis functions in, without consideration of which ones interact with each other. This method is the easiest to implement on a computer, but teaches us little about how the bonding in a molecule is built up. As we will now see, we can learn a lot more if we take this in a step-by-step fashion.

SYMMETRIZED BASIS FUNCTIONS

For N$_2$, our basis set consists of 2s, 2p$_x$, 2p$_y$, and 2p$_z$ orbitals centered on each of the N atoms. From the conventions we have been using, the possible wavefunctions are expressed in terms of a vector of coefficients as follows:

$$\psi = \begin{pmatrix} c\left(N1\ 2s\right) \\ c\left(N1\ 2p_x\right) \\ c\left(N1\ 2p_y\right) \\ c\left(N1\ 2p_z\right) \\ c\left(N2\ 2s\right) \\ c\left(N2\ 2p_x\right) \\ c\left(N2\ 2p_y\right) \\ c\left(N2\ 2p_z\right) \end{pmatrix}$$

The atomic orbitals then are expressed as vectors with one coefficient being one and all others zero.

The form of the Hamiltonian matrix can be greatly simplified if we recognize that in quantum mechanics pairs of symmetrically equivalent basis functions generally appear as symmetric or antisymmetric combinations. For instance, the N 2s orbitals will not appear separately in a MO, but instead as sums and differences. We can then replace our individual N_2 atomic orbitals in MATLAB with such linear combinations, e.g.:

```
>> psi_s_symm = sqrt(0.5)*[1 0 0 0 1 0 0 0]';
%   Prime: matrix transpose
>> psi_s_anti = sqrt(0.5)*[1 0 0 0 -1 0 0 0]';
>> psi_pz_symm = sqrt(0.5)*[0 0 0 1 0 0 0 -1]';
  % Note the sign change from psi_s_symm
>> psi_pz_anti = sqrt(0.5)*[0 0 0 1 0 0 0 1]';
```

where the sqrt(0.5) factor serves to normalize the functions. Matrix elements for a new Hamiltonian based on these symmetrized functions can then be calculated:

```
>> H_SALC(1,1) = psi_s_symm'*H_N2*psi_s_symm;
>> H_SALC(1,2) = psi_s_symm'*H_N2*psi_s_anti;
>> H_SALC(1,3) = psi_s_symm'*H_N2*psi_pz_symm;
>> H_SALC(1,4) = psi_s_symm'*H_N2*psi_pz_anti;
...
```

```
>> H_SALC(2,1) = psi_s_anti'*H_N2*psi_s_symm;
>> H_SALC(2,2) = psi_s_anti'*H_N2*psi_s_anti;
>> H_SALC(2,1) = psi_s_anti'*H_N2*psi_pz_symm;
>> H_SALC(2,2) = psi_s_anti'*H_N2*psi_pz_anti;
...
```

where "SALC" is an abbreviation for symmetry-adapted linear combinations, a term we will discuss in more detail as we get deeper into group theory in later chapters.

Exercise 5.4. Create the full set of symmetrized basis functions, and finish generating the 16×16 Hamiltonian matrix for this basis. Also, plot pictures of the symmetrized orbitals and paste screenshots of them on a presentation slide. Label each function in the slide with its energy (as can be read from the diagonal elements of the H_SALC matrix). Examples are shown in Figure 5.2.

Once the H_SALC matrix is completed, you should find that a vast majority of the off-diagonal matrix elements is zero (at least to the precision with which our overlap integrals were calculated), and that some do not interact with any other functions. These non-interacting functions are already completed MOs! For those sets that do interact, we can make mini subsystem Hamiltonians to resolve the bonding between them. For

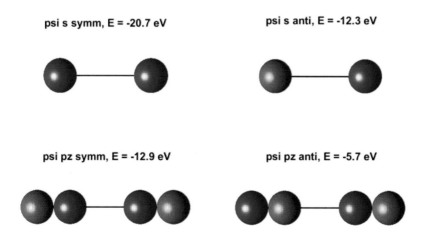

FIGURE 5.2 Examples of symmetry-adapted linear combinations of the valence atomic orbitals for the N atoms in an N_2 molecule.

instance, the symmetric combination of s orbitals interacts only with the symmetric combination of the p_z orbitals:

```
>> H_MINI(1,1) = psi_s_symm'*H_N2 *psi_s_symm;
>> H_MINI(1,2) = psi_s_symm'*H_N2 *psi_pz_symm;
>> H_MINI(2,1) = psi_pz_symm'*H_N2 *psi_s_symm;
>> H_MINI(2,2) = psi_pz_symm'*H_N2 *psi_pz_symm;
```

The subsystem Hamiltonian can then be diagonalized, to given eigenenergies and the eigenvectors in the basis of the s_symm and pz_symm functions:

```
>> [vect, E] = eig(H_MINI);
```

The eigenvectors can then be converted back to the atomic orbital basis:

```
>> psi1 = vect(1,1)*psi_s_symm + ...
   vect(2,1)*psi_pz_symm;
>> psi2 = vect(1,2)*psi_s_symm + ...
   vect(2,2)*psi_pz_symm;
```

And the psi1 and psi2 vectors can be plotted as any other MOs.

Exercise 5.5. On a presentation slide, create an MO diagram for the interaction of psi_s_symm and psi_pz_symm to create psi1 and psi2. Include plots of all four functions as well as their energies. Interpret the interaction, explaining the relative energies of the four orbitals. Your final plot should look something like Figure 5.3.

Here, the bonding combination of s orbitals has incorporated a little of the bonding combination of p orbitals, to slightly hybridize the functions on the two N atoms inwards along the bond. To maintain orthogonality, the p-based orbital has had to add the s-based one with the opposite sign, leading to the hybridization of the function away from the bond. A look at the coefficients in the vect array shows that this mixing is of exactly equal magnitude. Overall, the result of this interaction is that the two original bonding functions are transformed into a strongly bonding orbital and an essentially nonbonding orbital.

Exercise 5.6. Repeat this process for any other subset of interacting orbitals among the symmetrized basis set. You should see that the s-s and p_z-p_z

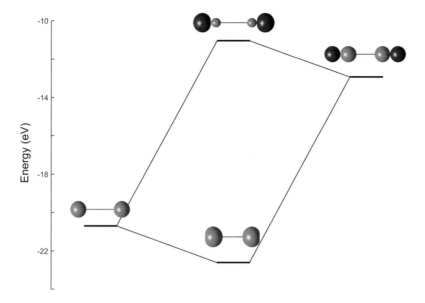

FIGURE 5.3   s-p hybridization between the bonding combinations of the N s and N p orbitals in N$_2$.

anti-bonding functions combine to make a nonbonding function and a strongly antibonding one.

Exercise 5.7. Show that the finalized MOs have been reached by taking the full set of functions that have been obtained and calculating a new Hamiltonian matrix for their interactions with each other. The result should be a diagonal matrix.

   Bonus code: When you are putting together MO diagrams, placing the energy levels one-by-one can be tiring and not necessarily subject to the same rigor as the rest of our analysis. Below you will find a MATLAB script for plotting these levels, which may be adapted for use in other exercises.

```
figure; hold on

% Set atomic orbital energies of starting atoms
% and MOs
AOenergies1 = [-16.5070 -9.3000 -9.3000 -9.3000]';
AOenergies2 = [-16.5070 -9.3000 -9.3000 -9.3000]';
  MOenergies = [   -22.5977
     -14.4025
```

```
        -11.0341
        -10.4590
        -10.4590
         -8.1410
         -8.1410
         -3.5797];

numAOs1 = size(AOenergies1);
numAOs1 = numAOs1(1);
numAOs2 = size(AOenergies2);
numAOs2 = numAOs2(1);

barwidth = 2;
deltaE_step = 0.3; % How far to space apart
                   % degenerate levels
deg_tol = 0.1; % How close two levels need to be
               % to be considered degenerate

step = 0;
for j=1:numAOs1
  deltaE=0;
  if(j>1) % code to stack up degenerate levels
    if(AOenergies1(j)-AOenergies1(j-1)<deg_tol)
      step = step + 1;
      deltaE = step*deltaE_step;
    else
      step = 0;
      deltaE = 0;
    end
  end
  plot([0,barwidth]+2*barwidth, ...
        [AOenergies1(j)+deltaE,AOenergies1(j)+ ...
        deltaE],'color',[0 0 0],'linewidth',2);
end

step = 0;
for j=1:numAOs2
  deltaE = 0
  if(j>1) % code to stack up degenerate levels
    if(AOenergies2(j) - AOenergies2(j-1)<deg_tol)
      step = step + 1;
      deltaE = step*deltaE_step;
```

```
    else
      step = 0;
      deltaE = 0;
    end
  end
  plot([barwidth*8,barwidth*8+barwidth]+2*barwidth,...
  [AOenergies2(j)+deltaE,AOenergies2(j)+deltaE],...
  'color',[0 0 0],'linewidth',2);
end

step = 0;
for j=1:(numAOs1+numAOs2)
  deltaE = 0
  if(j>1) % code to stack up degenerate levels
    if(MOenergies(j) - MOenergies(j-1)<deg_tol)
      step = step + 1;
      deltaE = step*deltaE_step;
    else
      step = 0;
      deltaE = 0;
    end
  end
  plot([barwidth*4,barwidth*4+barwidth]+...
  2*barwidth,[MOenergies(j)+deltaE,MOenergies(j)...
  +deltaE],
  'color',[0 0 0],'linewidth',2);
end
axis([2,22,-24,-3]) % Adjust axis limits here.
```

CHAPTER 6

Heteronuclear Diatomic Molecules

INTRODUCTION

We are now ready to move up one level of complexity by considering diatomic molecules built from atoms of two different elements. While such molecules are still quite simple compared to the most encountered in inorganic or organic chemistry, they offer some important lessons on how electronegativity differences affect bonding and reactivity. As we will see, they also provide an entry way into the use of perturbation theory in solving quantum mechanical problems, and yield the basic MO diagram for a Lewis acid-base interaction.

DRAWING MOLECULAR STRUCTURES

As we begin considering MO diagrams for molecules with more than one element, we can quickly run into confusion. From the orbital lobes alone it is often difficult to figure out which atom is which. For this reason, it is a good investment of time to write a few basic molecular graphics functions. Let's start with a simple one for creating wire-diagrams, in which the elemental identities are encoded with color. As before, we assume that the molecular geometry will be supplied in an XYZ file that can be read with `textread`:

```
[atomname,x,y,z] = textread(filename,'%s %f %f %f');
```

Then, we loop through the pairs of atoms and test whether the inter-atomic distance, dist, for each pair is within a specified range of dmin ≤ dist ≤ dmax, and draw a line between the two atoms whenever this condition is satisfied:

```
natoms = size(x);
natoms = natoms(1);

for j = 1:natoms
 for k = 1:natoms
    dist = ( (x(j)-x(k))^2 + (y(j)-y(k))^2 + ...
             (z(j)-z(k))^2 )^0.5;
    if( (dist<=d_max) && (dist>=d_min) )
       plot3([x(j) x(k)],[y(j) y(k)], [z(j) z(k)]);
    end
  end
end
```

This code by itself would give us the basic framework of the molecule, but not information about which atom is which element. It would be better if we could draw the line of the bond in a "bicolor" fashion, in which the line is divided into two segments which are color coded for which element that segment is connected to. To do this, we simply add some more conditions for when a bond is going to be drawn, and draw the bond in two steps: From atom1 to the midpoint of the bond with color1, and from the midpoint to atom2 with color2, as is shown here:

```
for j = 1:natoms
 for k = 1:natoms
  if(strcmp(atom1,atomname(j))==1)
   if(strcmp(atom2,atomname(k))==1)
    dist = ( (x(j)-x(k))^2 + (y(j)-y(k))^2 + ...
             (z(j)-z(k))^2 )^0.5;
    if( (dist<=d_max) && (dist>=d_min) )
    hold on;
    midpoint = [(x(j)+x(k))/2 (y(j)+y(k))/2 ...
                (z(j)+z(k))/2];
    % Draw first half of bond with color1
    plot3([x(j) midpoint(1)],[y(j) midpoint(2)],...
          [z(j) midpoint(3)],...
          'color',color1,'linewidth',width);
    % Draw second half of bond with color2
```

```
      plot3([midpoint(1) x(k)],[midpoint(2) y(k)], ...
            [midpoint(3) z(k)],...
            'color',color2,'linewidth',width);
    end
   end
  end
 end
end
```

The full code for the function is then

```
function draw_bonds(filename,atom1,color1,atom2,...
  color2,d_min,d_max,width)

[atomname,x,y,z] = textread(filename,'%s %f %f %f');

natoms = size(x);
natoms = natoms(1);

for j = 1:natoms
 for k = 1:natoms
  if(strcmp(atom1,atomname(j))==1)
   if(strcmp(atom2,atomname(k))==1)
    dist = ( (x(j)-x(k))^2 + (y(j)-y(k))^2 + (z(j)-...
             z(k))^2 )^0.5;
    if( (dist<=d_max) && (dist>=d_min) )
     % TIME TO DRAW A BOND
     hold on;
     midpoint = [(x(j)+x(k))/2 (y(j)+y(k))/2 ...
                 (z(j)+z(k))/2];
     % Draw first half of bond with color1
     plot3([x(j) midpoint(1)],[y(j) midpoint(2)], ...
           [z(j) midpoint(3)],'color',color1,...
           'linewidth',width);
     % Draw second half of bond with color2
     plot3([midpoint(1) x(k)],[midpoint(2) y(k)], ...
           [midpoint(3) z(k)], 'color',color2,...
           'linewidth',width);
    end
   end
  end
 end
end
```

From examining the above function, you can see that it draws all the bonds within a molecule for a given pair of elements specified by the variables `atom1` and `atom2`. This function can then be repeated to draw all of the desired bonds for the molecule.

In addition to this wire representation, we also have most of the programming we need to make ball-and-stick images. When we were analyzing the bonding in H_2 in Chapter 3, we made the function `draw_molecule_spheres` for plotting spheres to represent nuclear positions. With a few modifications, this can be adapted to give atoms of different elements their own colors and sphere radii:

```
function draw_spheres(filename,atom1,color,...
  sphere_radius,sphere_resolution);

[atomname,x,y,z] = textread(filename,'%s %f %f %f');

natoms = size(x);
natoms = natoms(1);
for j = 1:natoms
 if(strcmp(atom1,atomname(j))==1)
    [x_sphere,y_sphere,z_sphere] = sphere...
      (sphere_resolution);
    [phi, theta_matlab, r] = cart2sph(...
      x_sphere,y_sphere,z_sphere);
    theta = pi/2 - theta_matlab;
    r_new = sphere_radius;
    [x_new, y_new, z_new] = ...
    sph2cart(phi,theta_matlab,r_new);
    x_new = x_new + x(j);
    y_new = y_new + y(j);
    z_new = z_new + z(j);
    surf(x_new,y_new,z_new,'FaceColor',color,...
      'EdgeColor','none');
  end
end

axis equal;
```

When this `draw_spheres` function is used in conjunction with `draw_bonds` function, rudimentary ball-and-stick models are created.

As a simple example, let's visualize the geometry of a hypothetical heteronuclear diatomic molecule whose bonding we will examine in the next section, HeH. For the geometry, we can just use for now a bond distance of 0.7 Å (a real experimental distance is hard to come by here), making the geometry file HeH-geo as follows:

```
He  0.0  0.0  0.0
H   0.7  0.0  0.0
```

We then type in the commands to run the draw_spheres and draw_bonds functions:

```
>> draw_spheres('HeH-geo','He',[1 0 0],.1,40)
>> draw_spheres('HeH-geo','H',[.8 .8 .8],.1,40)
>> light
>> draw_bonds('HeH-geo','He',[1 0 0],'H',...
              [.8 .8 .8], 0.1,3.2,6)
>> axis off
```

which should result in a plot similar to that in Figure 6.1 (aside from the annotations, which take a little extra work using the menus of the figure window).

Exercise 6.1. **Plotting molecular geometries.** Use the functions created above to plot the structure of the hypothetical HeH molecule as a well as a handful of small molecules containing at least three atoms (Wikipedia is a good source of geometrical information from which Cartesian coordinates can be derived).

HeH: ELECTRONEGATIVITY PERTURBATION

Many of the changes in electronic structure that occur on going from a homonuclear diatomic molecule to a heteronuclear one can be illustrated

He   H

FIGURE 6.1   Image of a HeH molecule generated using the draw_spheres and draw_bonds functions.

with a simple model system, a hypothetical molecule consisting of a He atom and a H one (or, alternatively, the experimentally observed HeH$^+$ cation). The valence atomic orbitals involved in the construction of the MOs are two 1s orbitals, just as for the MOs of H$_2$. The major difference is that the He 1s and H 1s orbitals have very different energies. An electron in an H 1s orbital has an ionization energy of 13.6 eV, while an electron in the 1s orbital of He$^+$ ion is four times greater, at 54.4 eV, due to the increased nuclear charge. From a Hückel point of view, this difference would be reflected as a difference in the H_{ii} values for the orbitals, while on a conceptual level we can think about it as the source of an electronegativity difference between H and He.

After all the work we did earlier on calculating the MOs of H$_2$, one might wonder if there is a way we could simply update the MOs of H$_2$ to account for this H_{ii} difference, rather than starting over. In fact, the formalism of perturbation theory in quantum mechanics provides just such a procedure. In perturbation theory, we start out with a system for which we know the eigenfunctions and eigenvalues of the Hamiltonian operator:

$$\hat{H}^{\circ}\left|\psi_j^{\circ}\right\rangle = E_j^{\circ}\left|\psi_j^{\circ}\right\rangle$$

For this particular case, $\hat{H}^{\circ}$ would be represented by the Hamiltonian matrix that we obtained before for H$_2$, and the $\left|\psi_j^{\circ}\right\rangle$'s would be represented by eigenvectors of this matrix:

```
>> H_o = [ -7.5280   -7.9760
           -7.9760   -7.5280 ];

>> [psi_o, E_o] = eig(H_o)
psi_o =

  0.7071   -0.7071
  0.7071    0.7071


E_o =

 -15.5040        0
        0   0.4480
```

Once this initial system is set up, we then consider a perturbation to the physical system through the addition of a (hopefully) small new term to the Hamiltonian operator:

$$\hat{H} = \hat{H}^{\circ} + \hat{H}'$$

For HeH, this perturbation could consist of a change to the H_{ii} value of one of the 1s orbitals. For example, an $\hat{H}'$ matrix that lowers the energy of orbital 1 by 5 eV would be:

```
>> H_pert = [ -5.00 0.0
               0.0  0.0 ];
```

So that the full Hamiltonian becomes:

```
>> H = H_o + H_pert

H =

  -12.5280  -7.9760
   -7.9760  -7.5280
```

When we have divided the full Hamiltonian operator in this way into a part for which the answers are known and a deviation, perturbation theory offers a recipe for calculating corrections to the original MOs and their energies to account for the influence of the perturbation. Each additional correction corresponds to a higher order of perturbation, i.e. $E_j = E_j^{\circ} + \Delta E_j^{\text{1st order}} + \Delta E_j^{\text{2nd order}} + \cdots$ and $|\psi_j\rangle = |\psi_j^{\circ}\rangle + 1\text{st order correction} + 2\text{nd order correction}\ldots$. For the energies, the expression up to the first two orders has a simple interpretation. These energies are given by:

$$E_j = E_j^{\circ} + \langle \psi_j^{\circ} | \hat{H}' | \psi_j^{\circ} \rangle + \sum_{k \neq j} \frac{\left| \langle \psi_k^{\circ} | \hat{H}' | \psi_j^{\circ} \rangle \right|^2}{E_j^{\circ} - E_k^{\circ}} + \cdots$$

in which the starting point is the original energy of the MO. This original energy is then modified by the first order correction, which is simply the expectation value for the perturbation operator using the original MOs. This term is just the energy change that would be expected if the wavefunctions remained fixed in the presence of the perturbation. The second term provides the first correction in which the wavefunction is considered as responding to the perturbation. It consists of sum over all other $|\psi_k^\circ\rangle$'s to account for the energetic effect of mixing some of each $|\psi_k^\circ\rangle$ into the $|\psi_j\rangle$ function in response to the perturbation. Notice that these changes are largest when the perturbation creates a large interaction between the two original functions, as is expressed in the matrix element $\langle\psi_k^\circ|\hat{H}'|\psi_j^\circ\rangle$, and becomes smaller as the energy difference between the two functions gets larger. This expression clearly breaks down when two degenerate functions are prompted to interact by a perturbation (where $E_j^\circ - E_k^\circ = 0$); in that case one must use the degenerate perturbation theory (which basically amounts to diagonalizing a 2×2 Hamiltonian matrix for the interaction of the degenerate functions).

The above corrections to the energy are easily implemented in MATLAB®. For the H_2 MOs, psi_o(:,1) and psi_o(:,2), the first order correction is calculated as:

```
>> delta_E_first_1 = psi_o(:,1)'*H_pert*psi_o(:,1)

delta_E_first_1 =

   -2.5000

>> delta_E_first_2 = psi_o(:,2)'*H_pert*psi_o(:,2)

delta_E_first_2 =

   -2.5000
```

Here, both orbitals are found to stabilized by 2.5 eV, i.e. half of the energy difference ΔH_{ii} that we used to lower the energy of atomic orbital 1 over atomic orbital 2. As in both MOs the electron density is evenly distributed over the two atomic orbitals, such that an electron spends half of its time on each, this stabilization by half of the ΔH_{ii} has a clear explanation.

The second order energy corrections can be similarly calculated:

```
>> delta_E_second_1 = (psi_o(:,2)'*H_pert*...
   psi_o(:,1))^2/(E_o(1,1)-E_o(2,2))

delta_E_second_1 =
```

```
    -0.3918

>> delta_E_second_2 = (psi_o(:,1)'*H_pert*...
   psi_o(:,2))^2/(E_o(2,2)-E_o(1,1))

delta_E_second_2 =

   0.3918
```

As is fitting to these being second order corrections, their values are significantly lower than the first order ones we just determined. Also notice that the shifts for the two orbitals are of equal magnitudes but of opposite signs. A look at the formula for the second order correction illustrates why. delta_E_second_1 and delta_E_second_2 differ only in the sign of the energy difference in the denominator. These signs conspire so that the lower energy of the two is always stabilized, while the higher energy one is destabilized by the same amount.

The overall energies of the MOs to the second order are then:

```
>> E_1 = E_o(1,1) + delta_E_first_1 + delta_E_second_1

E_1 =

   -18.3958

>> E_2 = E_o(2,2) + delta_E_first_2 + delta_E_second_2
E_2 =

   -1.6602
```

The full evolution of the MO energies under these perturbations is given in Figure 6.2.

How did the interaction between MOs 1 and 2 give rise to these second order corrections to the energy? The answer to this question is found in the perturbation theory's prescription for calculating how a wavefunction evolves under a perturbation. To the first order this recipe is:

$$\left|\psi_j\right\rangle = \left|\psi_j^\circ\right\rangle + \sum_{k \neq j} \left(\frac{\left\langle \psi_k^\circ \middle| \hat{H}' \middle| \psi_j^\circ \right\rangle}{E_j^\circ - E_k^\circ} \right) \left|\psi_k^\circ\right\rangle + \cdots$$

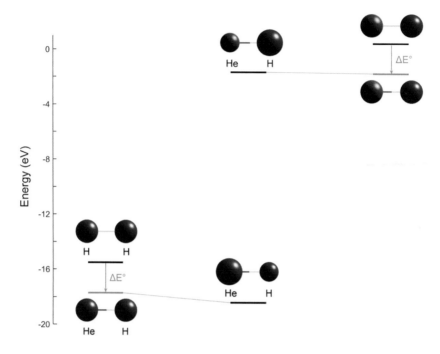

FIGURE 6.2   Construction of the MO diagram of HeH through perturbations on that of H_2.

where the change to the wavefunction is created by adding bits of the other original eigenfunctions (reflecting the postulate that the eigenfunctions of a quantum mechanical operator form a complete set). The coefficient for each of the $\left|\psi_k^\circ\right\rangle$'s is very similar to the second order corrections to the energy. The essential difference is that while the energetic corrections involve the squared magnitude of the interaction of the two functions, $\left|\left\langle\psi_k^\circ\middle|\hat{H}'\middle|\psi_j^\circ\right\rangle\right|^2$, the wavefunction correction is proportional to $\left\langle\psi_k^\circ\middle|\hat{H}'\middle|\psi_j^\circ\right\rangle$ without the squaring. The sign of $\left\langle\psi_k^\circ\middle|\hat{H}'\middle|\psi_j^\circ\right\rangle$ then becomes important in determining how the $\left|\psi_k^\circ\right\rangle$ is mixed into $\left|\psi_j^\circ\right\rangle$.

The MATLAB code for calculating to the changes to the wavefunctions is similar to that which we used for the energies:

```
psi_1 = psi_o(:,1) + (psi_o(:,2)'*H_pert*...
        psi_o(:,1))/(E_o(1,1)-E_o(2,2))*psi_o(:,2);

psi_2 = psi_o(:,2) + (psi_o(:,1)'*H_pert*...
        psi_o(:,2))/(E_o(2,2)-E_o(1,1))*psi_o(:,1);
```

The perturbed MOs are shown alongside the corrected energy levels in the HeH MO diagram above. Here, the lower energy bonding function has incorporated a little of -psi _ o(:,2)which has the stabilizing effect of shifting the electron density of the MO toward the more electronegative He atom. The antibonding function undergoes the reverse: It adds in the same amount of +psi _ o(:,1) to maintain orthogonality to the bonding MO. The result is a concentration of the electron density onto the less electronegative H atom. These changes are in accord with the variational principle, as it minimizes the energy of the lowest energy MO, even if it means destabilizing a higher energy one.

Exercise 6.2. **The HeH MO diagram.** Recreate the MO diagram of HeH shown in Figure 6.2 using the MATLAB-based implementation of perturbation theory illustrated above.

Exercise 6.3. **The HeH MO diagram without perturbation theory.** Compare the results that you obtained in Exercise 6.2 with the MOs and energy eigenvalues that you obtain from directly diagonalizing the H = H _ o + H _ pert matrix. How well does perturbation theory do at reproducing this "exact" solution? Are there any qualitative differences in the shapes of the MOs?

HeH: INTERATOMIC INTERACTIONS AS A PERTURBATION

Perturbation theory works best when the magnitude of the perturbation is very small. This need for a small perturbation is the reason we moved down the H_{ii} value of the He atom in the last section by only 5 eV, rather than the more appropriate 40.8 eV that one would get by comparing the ionization energies of H and He⁺. Such a large H_{ii} difference would in fact be much bigger than the interaction strength between the atomic orbitals on the two atoms. In this case, it would actually be a better breakdown of the Hamiltonian matrix to consider the two atomic orbitals as originally independent and treat their interaction as a perturbation.

In terms of the Hamiltonian matrix, we would then define the initial system and perturbation as:

```
>> H_o = [ -12.5280   0
               0      -7.5280 ];
>> H_pert = [ 0    -3
             -3     0 ];
```

```
>> [psi_o, E_o] = eig(H_o)
psi_o =
    1    0
    0    1


E_o =

 -12.5280        0
       0  -7.5280
```

where we've kept the same energy difference between the He and H H_{ii}'s, but decreased the magnitude of the interaction between the orbitals. The initial eigenvectors then consist of the individual He and H atomic orbitals, whose energies are just the H_{ii} values for those orbitals.

We next calculate the first and second order corrections to the atomic orbital energies due to their interaction. The first order terms are given by:

```
>> delta_E_first_1 = psi_o(:,1)'*H_pert*psi_o(:,1)

delta_E_first_1 =

    0

>> delta_E_first_2 = psi_o(:,2)'*H_pert*psi_o(:,2)

delta_E_first_2 =

    0
```

The zeros here arise because the perturbation matrix has no diagonal elements. The energetic effects of the perturbation are entirely due to the interactions between orbitals. These interactions are the domain of second order perturbation theory, as is given here:

```
>> delta_E_second_1 = (psi_o(:,2)'*H_pert*...
    psi_o(:,1))^2/(E_o(1,1)-E_o(2,2))

delta_E_second_1 =

   -1.8000
```

```
>> delta_E_second_2 = (psi_o(:,1)'*H_pert*...
    psi_o(:,2))^2/(E_o(2,2)-E_o(1,1))

delta_E_second_2 =

   1.8000
```

Again, the lower energy function is seen to be stabilized by the interaction, while the higher energy one is destabilized by the same amount.

The nature of these interactions is evident in the forms of the corrected wavefunctions:

```
>> psi_1 = psi_o(:,1) + (psi_o(:,2)'*H_pert*...
    psi_o(:,1))/(E_o(1,1)-E_o(2,2))*psi_o(:,2)

psi_1 =

   1.0000
   0.6000

>> psi_2 = psi_o(:,2) + (psi_o(:,1)'*H_pert*...
    psi_o(:,2))/(E_o(2,2)-E_o(1,1))*psi_o(:,1)

psi_2 =

  -0.6000
   1.0000
```

These results are summarized in the interaction diagram of Figure 6.3. The He atomic orbital is stabilized by the inclusion of a small amount of the H atomic orbital in an in-phase bonding fashion. This interaction results in a bonding MO polarized towards the more electronegative side of the molecule. In accordance with the variational principle, this MO is maximally stable in that it is both bonding and focused on the more electronegative atom. The higher energy H atomic orbital is destabilized in the reverse process: It subtracts the same amount of the He 1s orbital to create an antibonding orbital polarized towards the less electronegative atom. Time devoted to meditating on this MO diagram will be well spent; it bears the essential features of all pair-wise interactions between two orbitals of different energies.

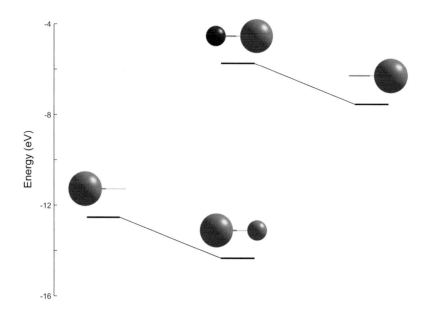

FIGURE 6.3   MO diagram for HeH derived from treating the interaction between the He 1s and H 1s orbitals as a perturbation. This diagram provides the basic framework for understanding Lewis acid-base interactions from the viewpoint of MO theory.

Exercise 6.4. **Another HeH MO diagram.** Create your own version of the above MO diagram for HeH. Populate it with the appropriate number of electrons for this molecule, and explain why HeH+ exists, while HeH does not.

Exercise 6.5. **Lewis acid-base interactions.** Adapt the above MO diagram to represent a simple model of a Lewis acid-base reaction. How would it need to be modified to represent, say, the creation of a Lewis acid-base adduct from NH_3 and BF_3?

THE MOs OF CO AND CN⁻

At this point, we have now extracted most of the insights available (at this level of theory) from a molecule based two s orbitals. Let's move on to some more involved examples of heteronuclear diatomic molecules, the isoelectronic CO and CN⁻. It turns out the same principles we discussed for HeH are at work in these molecules as well. In fact, these principles clearly explain why these species make such powerful ligands. In our analysis, we'll focus on the latter, as we already have nice parameters for C and N included in our `build_hamiltonian` function.

Exercise 6.6. **The MO diagram of CN⁻.** Consider a CN⁻ anion in which the N atom is at the origin, and the C atom is 1.15 Å away along the z-axis. Perform a simple Hückel calculation on this molecule using the MATLAB functions that we have built up to this point. Plot the resulting MO diagram, including both pictures of the MOs and the MO energies, and populate the diagram with the 10 valence electrons of the molecule. The result should resemble the plot in Figure 6.4.

Exercise 6.7. **The frontier orbitals of CN⁻.** The reactivity of molecules is often dictated by the shapes, nodal properties, and energies of their highest occupied molecular orbitals (HOMOs) and lowest unoccupied molecular orbitals (LUMOs), or other MOs nearby in energy. In the case of CN⁻ (and CO, which has MOs that are qualitatively the same as those of CN⁻), the HOMO and LUMO are both concentrated on the C atom, as

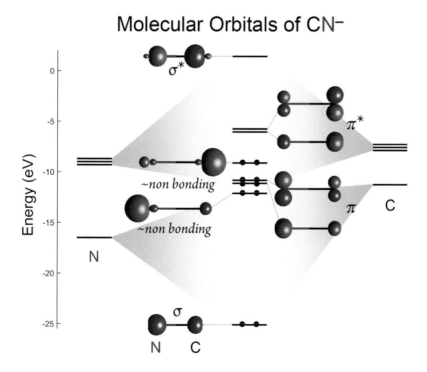

FIGURE 6.4 Hückel MO diagram of the CN⁻ ligand.

Frontier Orbitals of CN⁻

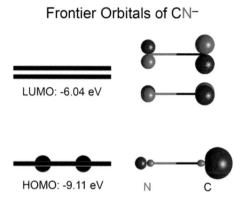

LUMO: -6.04 eV

HOMO: -9.11 eV

N   C

FIGURE 6.5   The frontier orbitals of the CN⁻ ligand.

shown in Figure 6.5. The σ character of the HOMO makes this molecule an excellent σ-donor on the C end, while the π* character of the LUMO makes it well-adapted for the molecule to serve as a π-acceptor at the C end. Explain using perturbation theory-based arguments why those MOs are polarized toward the C end of the molecule.

CHAPTER 7

Symmetry Operations

INTRODUCTION

So far in our bonding analyses, we have considered relatively simple diatomic molecules. As we go further, the Hamiltonian matrices that we work with will become increasing large, making attempts to understand them all at once extremely difficult. One aspect of molecular geometry that is extremely helpful here is symmetry, as it places strict conditions on the wavefunctions of a system. The mathematical formalism of group theory allows us to make use of symmetry to the fullest extent possible. This and the next couple chapters will thus focus on group theory and its expression in MATLAB® programming. Here, we will take the first step by considering symmetry operations.

APPLYING SYMMETRY OPERATIONS IN MATLAB

Symmetry operations are geometrical transformations on an object that result in configurations indistinguishable from the original. Examples can include rotations about special axes, reflections across planes, or inversion through a point. The geometrical elements about which these operations are defined are referred to as *symmetry elements*. Symmetry elements are points, lines, and planes for inversions, rotations, and reflections, respectively.

For molecular symmetry, operations can invariably be written as matrices which transform one point (x, y, z) into another point (x', y', z'):

$$
\begin{pmatrix} x' \\ y' \\ z' \end{pmatrix} = \begin{pmatrix} R_{11} & R_{12} & R_{13} \\ R_{21} & R_{22} & R_{23} \\ R_{31} & R_{32} & R_{33} \end{pmatrix} \begin{pmatrix} x \\ y \\ z \end{pmatrix} = \hat{R} \begin{pmatrix} x \\ y \\ z \end{pmatrix}
$$

It is then helpful to begin with creating a MATLAB function that can read in a list of atomic positions, just as our build_hamiltonian function does, and output a transformed list of positions. The header for this function could have the form

```
function apply_operation(filename1, filename2, ...
    Rmatrix)
```

where filename1 and filename2 are the names of the files containing the input and output geometries, and Rmatrix is a 3 × 3 matrix describing the symmetry operation.

Within this function, we start by reading in the initial geometry using lines familiar from build_hamiltonian:

```
[name,x,y,z] = textread(filename1,'%s %f %f %f');
atomnum = size(x);
atomnum = atomnum(1);
```

We next loop over the atomic positions, transforming the original x, y, z positions to the new positions x_new, y_new, z_new:

```
for j = 1:atomnum
  xyz_old = [x(j) y(j) z(j)]';
  xyz_new = Rmatrix*xyz_old;
  x_new(j,1) = xyz_new(1,1);
  y_new(j,1) = xyz_new(2,1);
  z_new(j,1) = xyz_new(3,1);
end
```

Now that we have the transformed atomic positions, we simply need to write them to the output file. First, we need to open the file:

```
f2 = fopen(filename2,'w');
```

Here, f2 is a handle that we will use whenever we work with the file, and 'w' specifies that we are opening the file in write mode (as opposed to the read or append modes). In this mode, MATLAB will create the file if it does not already exist, or overwrite an existing file. Once the file is opened, we can then loop over the transformed atomic positions and write them to the file, and finally close the file:

```
for j = 1:atomnum
  fprintf(  '%s %f %f %f -> ',char(name(j)),...
    x(j,1),y(j,1),z(j,1));
  % char(name(j)) converts name(j) from the cell
  % format it is given by the textread
  % function to the normal string format.
  fprintf(  '%s %f %f %f \n',char(name(j)),...
    x_new(j,1),y_new(j,1),z_new(j,1));
  fprintf(f2,'%s %f %f %f \n',char(name(j)),...
    x_new(j,1),y_new(j,1),z_new(j,1));
end
fclose(f2);
```

Note that the first two `fprintf` commands within the loop do not mention the `f2` handle; this text gets printed to the screen to show how the original atomic positions are transformed to the new positions. The final line outputs the atomic position in our standard format to the file. You will notice the sequence "\n" in the `fprintf` lines. This is a special character combination instructing the function to start a new line.

We are now ready to put the whole function together:

```
function apply_operation(filename1, filename2,...
  Rmatrix)

[name,x,y,z] = textread(filename1,'%s %f %f %f');
name = char(name);
atomnum = size(x);
atomnum = atomnum(1);

for j = 1:atomnum
  xyz_old = [x(j) y(j) z(j)]';
  xyz_new = Rmatrix*xyz_old;
  x_new(j,1) = xyz_new(1,1);
  y_new(j,1) = xyz_new(2,1);
  z_new(j,1) = xyz_new(3,1);
end

f2 = fopen(filename2,'w');

for j = 1:atomnum
  fprintf('%s %f %f %f -> ',name(j),x(j,1),...
    y(j,1),z(j,1));
```

```
    fprintf('%s %f %f %f\n',name(j),x_new(j,1),...
        y_new(j,1),z_new(j,1));
    fprintf(f2,'%s %f %f %f\n',name(j),x_new(j,1),...
        y_new(j,1),z_new(j,1));
end
fclose(f2);
```

As we consider the symmetry operations that can be applied with this function, it is nice to have a specific molecule to use as an example. We'll use a tetrahedral molecule similar to CH_4, but with the H atoms renamed to keep track of which atom is which:

```
C   0.000000   0.000000   0.000000
H   1.414214   0.000000   1.000000
Cl -1.414214   0.000000   1.000000
Br  0.000000 -1.414214 -1.000000
F   0.000000   1.414214 -1.000000
```

Exercise 7.1. **Plotting our starting point.** Write a function which takes an input geometry file and draws a picture of the molecule using the draw_bonds and draw_spheres functions. Explicitly set the view and camup directions so that several pictures can be created that are all seen from the same viewpoint. This consistency will be important for seeing how symmetry operations transform a molecule. Use this function to portray the above tetrahedral molecule. The result should be similar to the image shown in Figure 7.1.

THE IDENTITY OPERATION

The first symmetry operation we consider is pretty simple, but necessary for the theorems of group theory: The identity operation, whose symbol in the Schönflies notation usually used for molecular symmetry is $\hat{E}$. This operation leaves the molecule unchanged, with each atomic position being mapped to itself:

$$\begin{pmatrix} x' \\ y' \\ z' \end{pmatrix} = \begin{pmatrix} 1 & 0 & 0 \\ 0 & 1 & 0 \\ 0 & 0 & 1 \end{pmatrix} \begin{pmatrix} x \\ y \\ z \end{pmatrix} = \hat{E} \begin{pmatrix} x \\ y \\ z \end{pmatrix} = \begin{pmatrix} x \\ y \\ z \end{pmatrix}$$

This operation plays the important role of ensuring that there is logical consistency in our symmetry analysis; if we have not done anything to the

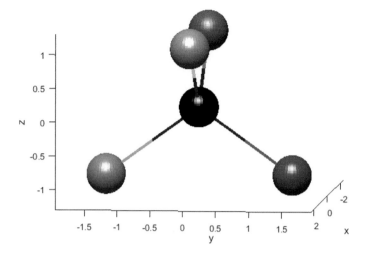

FIGURE 7.1   A tetrahedral molecule with the four corners marked with different colors, to help keep track of them as we apply symmetry operations.

molecule, its configuration should not have changed (in the hypothetical world of static molecules that endure whatever symmetry operations we wish to perform on them).

Exercise 7.2. **Debugging code with the identity operation.** This is our first chance to test out our `apply_operation` function. Use it to apply the identity operation to our sample molecule, and show that it results in an identical configuration to the starting point (the positions of the colors are the same). If the resulting molecule looks different from the original, it is time to track down some bugs in the function.

INVERSION THROUGH A CENTRAL POINT

The next symmetry operation to consider is an inversion through a point, such that every atomic position is transformed to the opposite position relative to the high symmetry point, which is known as an inversion center. When the inversion center is at the origin, the transformation is simply $(x, y, z) \rightarrow (-x, -y, -z)$. In matrix notation, this has the form:

$$\begin{pmatrix} x' \\ y' \\ z' \end{pmatrix} = \begin{pmatrix} -1 & 0 & 0 \\ 0 & -1 & 0 \\ 0 & 0 & -1 \end{pmatrix} \begin{pmatrix} x \\ y \\ z \end{pmatrix} = \hat{i} \begin{pmatrix} x \\ y \\ z \end{pmatrix} = \begin{pmatrix} -x \\ -y \\ -z \end{pmatrix}$$

where $\hat{i}$ is the Schönflies symbol for the inversion operation.

Exercise 7.3. **Inverting a tetrahedron.** Show that applying the inversion operation to our tetrahedral molecule yields the configuration shown in Figure 7.2. Note that every terminal atom has passed through the central atom to take a position on the opposite side with a distance equal to that of its original position. These new positions were not occupied by atoms before the inversion was performed, indicating that this is a distinguishable configuration from the original. A tetrahedron thus does not have inversion symmetry, unlike, say, an octahedral or square planar geometry.

REFLECTIONS THROUGH A PLANE

Reflections transform every point on one side of a plane into its mirror image on the other side of the plane. The plane through which this reflection is carried out is a mirror plane, and is denoted by σ in the Schönflies notation. For planes that run perpendicular to the axes of a Cartesian coordinate system and pass through the origin, these operations have a simple form. They just take each atom's coordinate along the axis perpendicular to the plane, and invert it. For example, a reflection through the xz-plane is given by:

$$\begin{pmatrix} x' \\ y' \\ z' \end{pmatrix} = \begin{pmatrix} 1 & 0 & 0 \\ 0 & -1 & 0 \\ 0 & 0 & 1 \end{pmatrix} \begin{pmatrix} x \\ y \\ z \end{pmatrix} = \hat{\sigma}_{xz} \begin{pmatrix} x \\ y \\ z \end{pmatrix} = \begin{pmatrix} x \\ -y \\ z \end{pmatrix}$$

The effect of this operation on a molecule is illustrated in Figure 7.3. The reflection through the xz-plane has inverted the y-coordinates of the atoms. For the black, gray, and green atoms the original y coordinates were zero, and the operation results in no change in their positions. The blue and orange atoms, however, face each other on opposite sides of the

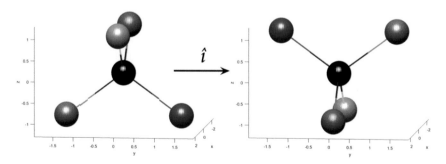

FIGURE 7.2 Application of an inversion operation to a tetrahedral molecule.

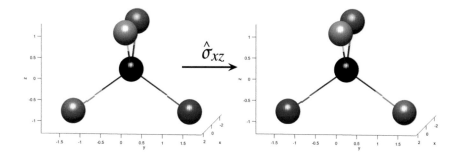

FIGURE 7.3   Application of a mirror reflection through the xz-plane on a tetra-hedral molecule.

xz-plane. Their positions are now interchanged. When we recall that our coloring of the atoms is simply to keep track of which one is which, rather than to denote different elements, we see that the molecule that results is equivalent to the original. $\hat{\sigma}_{xz}$ is thus a symmetry operation for this molecule.

<u>*Exercise 7.4.*</u> **More mirror operations.** Show that $\hat{\sigma}_{yz}$ is also a symmetry operation for the molecule, but that (due to the peculiarities of the orientation of the molecule in the coordinate system) $\hat{\sigma}_{xy}$ is not.

ROTATIONS ABOUT AN AXIS

Molecules can also exhibit rotational symmetry, in which rotations in certain increments around a special axis leave the molecule in a configuration indistinguishable from the original. The Schönflies notation for these rotational operations is $\hat{C}_n$, where n is the number of these operations that are necessary to complete a 360° turn, i.e. each operation turns the molecule by 360°/n. For the case of the rotations being around an axis of the coordinate system, such as z, the effect of the rotation on the atomic positions is simple to work out using trigonometry.

Let's start with a rotation around the z-axis by $\Delta\phi$. As the z-coordinates will be unaffected by the operation, we can focus on the changes to x and y. The first step is to write x and y in polar coordinates.

$$x = r\cos(\phi)$$

$$y = r\sin(\phi)$$

The rotation in this coordinate system consists simply of adding $\Delta\phi$ to the initial value of ϕ:

$$x' = r \cos(\phi + \Delta\phi)$$

$$y' = r \sin(\phi + \Delta\phi)$$

These transformed coordinates can then be expressed as linear combinations of the old coordinates through the following trigonometric identities (which are conveniently listed on Wikipedia for those of us who have trouble remembering them):

$$\cos(\phi + \Delta\phi) = \cos(\phi)\cos(\Delta\phi) - \sin(\phi)\sin(\Delta\phi)$$

$$\sin(\phi + \Delta\phi) = \cos(\phi)\sin(\Delta\phi) + \sin(\phi)\cos(\Delta\phi)$$

Substituting these identities into our expressions for x' and y' gives:

$$x' = r \cos(\phi + \Delta\phi) = r \cos(\phi)\cos(\Delta\phi) - r \sin(\phi)\sin(\Delta\phi)$$

$$= x \cos(\Delta\phi) - y \sin(\Delta\phi)$$

$$y' = r \sin(\phi + \Delta\phi) = r \cos(\phi)\sin(\Delta\phi) + r \sin(\phi)\cos(\Delta\phi)$$

$$= x \sin(\Delta\phi) + y \cos(\Delta\phi)$$

which can be written in matrix form as:

$$\begin{pmatrix} x' \\ y' \end{pmatrix} = \begin{pmatrix} \cos(\Delta\phi) & -\sin(\Delta\phi) \\ \sin(\Delta\phi) & \cos(\Delta\phi) \end{pmatrix} \begin{pmatrix} x \\ y \end{pmatrix}$$

or for three dimensions as:

$$\begin{pmatrix} x' \\ y' \\ z' \end{pmatrix} = \begin{pmatrix} \cos(\Delta\phi) & -\sin(\Delta\phi) & 0 \\ \sin(\Delta\phi) & \cos(\Delta\phi) & 0 \\ 0 & 0 & 1 \end{pmatrix} \begin{pmatrix} x \\ y \\ z \end{pmatrix} = \hat{C}_{\frac{360°}{\Delta\phi}} \begin{pmatrix} x \\ y \\ z \end{pmatrix}$$

Let's try this out on our sample molecule. To set up a 180-degree rotation around the z-axis we first create the transformation matrix in MATLAB:

```
>> delta_phi = 180;
>> C_2 = [cosd(delta_phi) -sind(delta_phi)  0
       sind(delta_phi) cosd(delta_phi)  0
            0          0      1]

C_2 =

    -1    0    0
     0   -1    0
     0    0    1
```

This operation can then be applied to the molecule to give the result displayed in Figure 7.4.

Exercise 7.5. **C₂ operations.** Confirm the above result, and show that while a C_2 rotation about the z-axis is a symmetry operation, the corresponding rotations about the x and y axes are not (for this orientation of the molecule).

IMPROPER ROTATIONS

The final distinct type of symmetry operation that can be applied to a molecule is an improper rotation. These are compound operations in which a rotation about an axis is followed by a reflection through a plane perpendicular to that axis. An example of such an operation is:

$$\begin{pmatrix} x' \\ y' \\ z' \end{pmatrix} = \begin{pmatrix} 1 & 0 & 0 \\ 0 & 1 & 0 \\ 0 & 0 & -1 \end{pmatrix} \begin{pmatrix} \cos(\Delta\phi) & -\sin(\Delta\phi) & 0 \\ \sin(\Delta\phi) & \cos(\Delta\phi) & 0 \\ 0 & 0 & 1 \end{pmatrix} \begin{pmatrix} x \\ y \\ z \end{pmatrix} = \hat{S}_{360°\atop \Delta\phi} \begin{pmatrix} x \\ y \\ z \end{pmatrix}$$

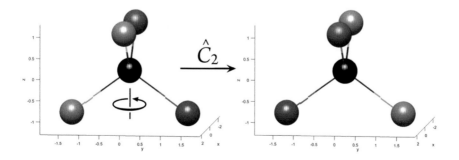

FIGURE 7.4 C_2 rotation of the tetrahedral molecule around the z-axis.

which consists of a rotation about z followed by a reflection through the xy-plane. In the Schönflies notation, these operations are denoted as $\hat{S}_n$, where n again refers to the number of operations that is necessary to complete a 360° turn.

A nice example of such an operation can be found in the tetrahedral sample molecule we have been using. If one were to look at the molecule down the z-axis, the terminal atoms would appear as the corners of a square in the projection. This square-like image hints at some form of four-fold symmetry, but the three-dimensional shape of the molecule deviates from a normal C_4 symmetry. However, when this C_4 operation is followed by a reflection, a true symmetry operation emerges (Figure 7.5).

Exercise 7.6. Recreate the Figure 7.5 showing how an $\hat{S}_4$ operation works, and show the 3×3 matrix for the overall operation.

CREATING MORE COMPLICATED OPERATIONS

Up until now, we have worked only with the simplest cases of symmetry operations, those whose symmetry elements are aligned with the axes of the coordinate system. Capturing the full symmetry of many molecules will require more work, as can be seen in our tetrahedral molecule. It has multiple three-fold axes, two-fold axes, and mirror planes that pass through the molecule at oblique angles to the axes. We do, however, already have the tools to deal with such cases if we are willing to combine a series of simple operations to make more complicated ones.

As an example, let's consider one of the three-fold symmetry axes of a tetrahedral molecule. In such a molecule, a three-fold axis is visible when we look along any of the bonds to the central atom. If we consider our line of sight as the z-axis, then a simple recipe for making a three-fold rotation emerges. When we hold a molecular model and decide to do a three-fold rotation, we first turn the molecule so that we look down the correct axis, then turn the molecule about that axis by 120°, and then reverse the motion we used to put the three-fold axis along z. This procedure is a specific example of a similarity transform, an important type of transformation we will return to in future chapters.

These steps are easy to implement in MATLAB. Consider the three-fold axis running along the bond connecting the gray atom to the central atom

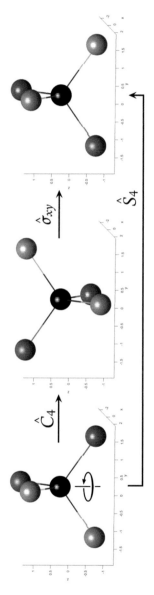

FIGURE 7.5   Application of an S_4 operation to a tetrahedral molecule.

in our sample molecule. The bond is tilted off of the z-axis in the xz-plane by half of the tetrahedral angle, $109.471°/2 = 54.7355°$. We can then reorient the molecule so that this bond is along the z-axis by rotating about y off of the x-axis by $54.7355°$:

```
>> delta_phi = 54.7355;
>> reorient = [cosd(delta_phi)  0  -sind(delta_phi)
                       0          1        0
               sind(delta_phi)  0  cosd(delta_phi)]

reorient =

   0.5774       0      -0.8165
      0      1.0000       0
   0.8165       0       0.5774
```

Applying this operation to the molecule has the effect shown in Figure 7.6, where the result is a molecule that has its black-gray bond oriented along z as desired.

Now, we can perform a $120°$ rotation about z using a matrix similar to those we have used before (Figure 7.7).

Finally, we put the black-gray bond back to its original position by reversing the $54.7355°$ rotation about the y-axis using the transformation matrix:

```
>> delta_phi = -54.7355;
>> reorient_inv = [cosd(delta_phi)  0  -sind(delta_phi)
                          0          1        0
                  sind(delta_phi)  0  cosd(delta_phi)]

reorient_inv =

   0.5774       0       0.8165
      0      1.0000       0
  -0.8165       0       0.5774
```

whose effect on the molecule is shown in Figure 7.8.

If we compare the end of this process with our original starting point, we see that we have indeed achieved a three-fold rotation about the black-gray bond (Figure 7.9).

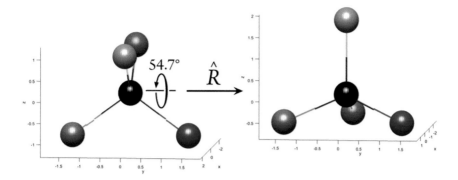

FIGURE 7.6 Rotation of tetrahedral molecule to place one bond along the z-axis.

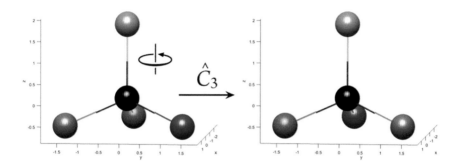

FIGURE 7.7 C_3 rotation of reoriented tetrahedral molecule around the z-axis.

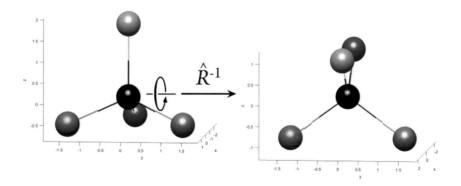

FIGURE 7.8 Reverse of the operation shown in Figure 7.6.

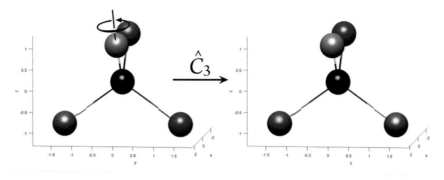

FIGURE 7.9 Rotation of a tetrahedral molecule around one of its C_3-axes.

The overall matrix for this process is given by:

```
>> C3 = reorient_inv*C_3_z*reorient

C3 =

    0.5000  -0.5000   0.7071
    0.5000  -0.5000  -0.7071
    0.7071   0.7071   0.0000
```

Matrices for other operations not aligned with a special axis or plane of the coordinate system can be similarly derived.

Exercise 7.7. Repeat the above process for the rotation of the tetrahedral molecule about another of its three-fold axes.

Exercise 7.8. Derive and demonstrate the action of a matrix for a C_2 rotation about a vector perpendicular to z and bisecting the x and y axes. Show that this is a symmetry operation for the tetrahedron.

CHAPTER 8

Symmetry Groups

INTRODUCTION

In the last chapter, we saw how the symmetry of molecules is reflected in the transformations that we can perform on them that result in a configuration indistinguishable from the original. The existence of these symmetry operations for a molecule has profound implications for the electronic structure and properties of a molecule. These implications become clear as we connect the symmetry operations of structures to the concept of a **mathematical group**.

PROPERTIES OF MATHEMATICAL GROUPS

A group, in the mathematical sense, is a collection of mathematical elements and a rule for taking products of them with each other that obey the following four rules:

Rule 1. The product of any member of the group is also a member of the group.

Rule 2. The group contains an element corresponding to an identity operation ($\hat{E}$).

Rule 3. The products of members are associative, e.g. $\left(\hat{A} \cdot \hat{B}\right) \cdot \hat{C} = \hat{A} \cdot \left(\hat{B} \cdot \hat{C}\right)$.

Rule 4. The inverse of each member of a group is also a member of the group, e.g. if $\hat{A}$ is a member, then so is $\hat{A}^{-1}$ such that $\hat{A}^{-1} \cdot \hat{A} = \hat{E}$.

As we will see in more detail below, the collection of symmetry operations (and the matrices used to represent them) that can be applied to any given molecule have these properties and can thus be considered to comprise a mathematical group. This connection between symmetry and mathematical groups allows us to apply the conclusions of group theory to the analysis of the electronic structure and geometries of molecules. For a molecule, its symmetry operations all leave the central point of the molecule invariant; molecular symmetry groups are thus known as **point groups**.

DEMONSTRATION OF MATHEMATICAL GROUPS WITH MATLAB

To see that molecular symmetry operations follow the above rules, let's consider a simple example: An H_2O molecule, oriented as shown in Figure 8.1. This molecule has a C_2-axis along the z-axis, and mirror planes in the xz and yz planes. Together with the $\hat{E}$ operation, this leads to group of four symmetry operations known as the point group C_{2v}. Using our work from the last chapter, we can easily set up matrices for these four operations in MATLAB®:

```
>> E = [1 0 0; 0 1 0; 0 0 1]

E =

    1    0    0
    0    1    0
    0    0    1

>> C2 = [-1 0 0; 0 -1 0; 0 0 1]

C2 =
```

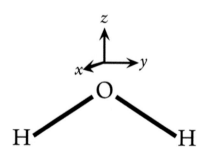

FIGURE 8.1   Water molecule with the coordinate system chosen for the discussion in the text.

```
-1   0   0
 0  -1   0
 0   0   1

>> sigma_xz = [1 0 0; 0 -1 0; 0 0 1]

sigma_xz =

 1   0   0
 0  -1   0
 0   0   1

>> sigma_yz = [-1 0 0; 0 1 0; 0 0 1]

sigma_yz =

-1   0   0
 0   1   0
 0   0   1
```

Once we have these matrices set up, we can then take products of them and try to identify the results. For example, doing two $\hat{C}_2$ operations in sequence leads to:

```
>> C2*C2

ans =

 1   0   0
 0   1   0
 0   0   1
```

which is of course the same as the identity operation. Likewise, performing $\hat{\sigma}_{yz}$ then $\hat{\sigma}_{xz}$ yields:

```
>> sigma_xz*sigma_yz

ans =

-1   0   0
 0  -1   0
 0   0   1
```

a matrix that is equal to that of a $\hat{C}_2$ rotation.

Through taking all of the possible products the following multiplication table can be developed:

	$\hat{E}$	$\hat{C}_2$	$\hat{\sigma}_{xz}$	$\hat{\sigma}_{yz}$
$\hat{E}$	$\hat{E}$	$\hat{C}_2$	$\hat{\sigma}_{xz}$	$\hat{\sigma}_{yz}$
$\hat{C}_2$	$\hat{C}_2$	$\hat{E}$	$\hat{\sigma}_{yz}$	$\hat{\sigma}_{xz}$
$\hat{\sigma}_{xz}$	$\hat{\sigma}_{xz}$	$\hat{\sigma}_{yz}$	$\hat{E}$	$\hat{C}_2$
$\hat{\sigma}_{yz}$	$\hat{\sigma}_{yz}$	$\hat{\sigma}_{xz}$	$\hat{C}_2$	$\hat{E}$

From an examination of this table, it can be confirmed that this collection of symmetry operations follows the rules for a mathematical group described above.

Exercise 8.1. **The multiplication table of the C_{3v} point group.** The point group of an NH_3 molecule is called C_{3v} and is analogous to that of H_2O but with the C_2-axis of H_2O replaced by a C_3-axis (see Figure 8.2). The full point group can be developed by applying beginning with just $\hat{E}$, a $\hat{C}_3$ operation about the z-axis, and a $\hat{\sigma}_{yz}$ operation, and then taking the products between these operations. Following this procedure, develop a set of

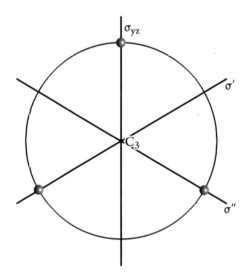

FIGURE 8.2   Symmetry elements in the C_{3v} point group.

matrices in MATLAB representing the operations of this point group, and build its multiplication table.

The C_{3v} point group can be used to illustrate some important aspects of point groups and symmetry operations. First, the order in which two operations are taken can affect the overall result. As an example, try comparing the matrices that result from the products $\hat{C}_3 \cdot \hat{\sigma}_{yz}$ and $\hat{\sigma}_{yz} \cdot \hat{C}_3$:

```
>> C_3*sigma_yz

ans =

    0.5000   -0.8660        0
   -0.8660   -0.5000        0
        0         0   1.0000

>> sigma_yz*C_3

ans =

    0.5000    0.8660        0
    0.8660   -0.5000        0
        0         0   1.0000
```

These two matrices are clearly distinct from each other, illustrating that in general symmetry operations cannot be expected to commute with each other.

The C_{3v} point group also demonstrates the concept of **classes** of symmetry operations. In the scheme above showing the symmetry elements in the C_{3v} point group, three mirror planes are present. These mirror planes are not entirely independent from each other, however. They are related to each other via the three-fold rotational symmetry of the system. As such, we can combine the matrix for one mirror operation, say that of the $\hat{\sigma}_{yz}$ operation, with the matrices for the $\hat{C}_3$ rotations to create matrices for the other mirror operations. This process simply uses the scheme for building complicated symmetry matrices from simpler ones in the last chapter. For example, to make a matrix for the $\hat{\sigma}''$ operation, we can rotate the σ'' plane onto the σ_{yz} plane through a $\hat{C}_3^{-1} = \hat{C}_3^{2}$ operation, perform the $\hat{\sigma}_{yz}$ operation, and then rotate the plane back to its original position using a $\hat{C}_3$ operation. The overall process is given by $\hat{\sigma}'' = \hat{C}_3 \cdot \hat{\sigma}_{yz} \cdot \hat{C}_3^{-1}$:

```
>> C_3_2 = C_3*C_3;
>> sigma_doubleprime = C_3*sigma_yz*C_3_2

sigma_doubleprime =

0.5000      0.8660           0
0.8660     -0.5000           0
     0           0      1.0000
```

Operations that are related to each other via other symmetry operations in the group in this way are said to belong to the same **class**. Formally, $\hat{A}$ and $\hat{A}'$ belong to the same class if they are connected through similarity transforms: $\hat{A}' = \hat{R}^{-1} \cdot \hat{A} \cdot \hat{R}$, where $\hat{R}$ and $\hat{R}^{-1}$ are members of the group.

Exercise 8.2. Show using matrix products in MATLAB that $\hat{C}_3$ and $\hat{C}_3^2$ in the C_{3v} point group belong to the same class.

GENERATING OPERATIONS

In our examples so far, we have seen how products of a small number of operations can give rise to a larger variety of new operations. In general, a select number of operators is sufficient to serve as a basis for the construction of the point group. This minimal group of operations necessary to reconstitute a group is known as the **generating operations** or **generators** for the group. We can greatly speed up the process of exploring point groups by creating a function that works through this process of building a group from its generators, which could take the form:

```
function FullGroup = BuildPointGroup(generators)
```

where generators is an array of matrices representing the generating operations.

How would this generators array look and be constructed? We can simply take the same matrix format we have been using until now, and add a third index to refer to different symmetry operations. In the C_{3v}

point group, for example, our generators are again $\hat{E}$, $\hat{C}_3$, and $\hat{\sigma}_{yz}$. These operators can be combined into a single array as follows:

```
>> C3v_generators(:,:,1) = E;
>> C3v_generators(:,:,2) = C_3;
>> C3v_generators(:,:,3) = sigma_yz;
```

The form of the resulting array can then be checked as follows

```
>> C3v_generators

C3v_generators(:,:,1) =

   1   0   0
   0   1   0
   0   0   1

C3v_generators(:,:,2) =

   -0.5000    -0.8660           0
   0.8660     -0.5000           0
   0              0      1.0000

C3v_generators(:,:,3) =

   -1   0   0
    0   1   0
    0   0   1
```

The first step in building a group for this list of generators is to figure out how many operations we currently have:

```
num_operations = size(generators);
num_operations = num_operations(3);
```

Then we use the generators array as the start for another array that will be expanded to include all other operations that we find through taking products:

```
operations = generators;
```

Finally, we cycle through loops of the taking products of all the operations we have thus far in the group, adding each new operation that arises in the process to our list:

```
stop = 0;
num_operations_start = num_operations;
loop_counter = 0;
while (stop == 0)
  stop = 1;
  for j = 1:num_operations_start
    for k = 1:num_operations_start
      new_matrix = operations(:,:,j)*operations(:,:,k);
      foundit = 0;
      for m = 1:num_operations
        if(norm(new_matrix-operations(:,:,m))<0.001)
          foundit = 1;
        end
      end
      if(foundit==0)
        stop = 0;
        num_operations = num_operations+1;
        operations(:,:,num_operations) = new_matrix;
      end
    end
  end
  num_operations_start = num_operations;
  loop_counter = loop_counter + 1;
  fprintf('Loop %d: total operations = %d\n',...
    loop_counter, num_operations);
end
```

The "while (stop == 0) ... end" loop repeats the enclosed lines as long as stop = 0 at the end of each cycle. Our first step in the loop is to change stop to 1, with the idea that it will only be switched to zero if we find a new operation during the cycle. This is followed by a double loop over the current list of operations to carry out all the possible products of the operations:

```
for j = 1:num_operations_start
  for k = 1:num_operations_start
    new_matrix = operations(:,:,j)*operations(:,:,k);
  end
end
```

As each product is taken, we need to check to see whether it represents a repeat of an operation we already have or is new. That is accomplished by taking the matrix norm of the difference between the new matrix and every operation in our list. If the norm of the difference is essentially zero (we use a small number here instead of zero to take care of possible round-off errors), then we have a repeat. If we cannot equate the new matrix to anything in our list, it is added to the list as a new operation:

```
foundit = 0;
for m = 1:num_operations
  if(norm(new_matrix-operations(:,:,m))<0.001)
    foundit = 1;
  end
end
if(foundit==0)
  stop = 0;
  num_operations = num_operations+1;
  operations(:,:,num_operations) = new_matrix;
end
```

When going through this cycle no longer leads to new operations, stop will stay equal to 1 until the end of the cycle. This breaks the while (stop == 0) condition, and the program will exit the loop. Here is the full function that combines all of these steps:

```
function FullGroup = BuildPointGroup(generators)

num_operations = size(generators);
num_operations = num_operations(3);

operations = generators;

stop = 0;
num_operations_start = num_operations;
loop_counter = 0;
while (stop == 0)
  stop = 1;
  for j = 1:num_operations_start
    for k = 1:num_operations_start
      new_matrix = operations(:,:,j)*operations...
        (:,:,k);
      foundit = 0;
```

```
        for m = 1:num_operations
            if(norm(new_matrix-operations(:,:,m))<0.001)
                foundit = 1;
            end
        end
        if(foundit==0)
            stop = 0;
            num_operations = num_operations+1;
            operations(:,:,num_operations) = new_matrix;
        end
      end
   end
   num_operations_start = num_operations;
   loop_counter = loop_counter + 1;
   fprintf('Loop %d: total operations = %d\n',...
      loop_counter, num_operations);
end

FullGroup = operations;
```

APPLYING GROUP OPERATIONS

Once we generate a full group of matrices representing the symmetry operations of a molecule, it becomes desirable to explore the symmetries they signify. To do this, let's write a function that will take the coordinates in a geometry file, and transform them using the operations in a group:

```
function apply_group(filename1, filename2, ...
operations)
```

The first steps in this function are to figure out the number of operations in the group (the order of the group) and import the geometry from the input file:

```
num_operations = size(operations);
num_operations = num_operations(3);

[name,x,y,z] = textread(filename1,'%s %f %f %f');
name = char(name);
name
```

```
atomnum = size(x);
atomnum = atomnum(1);
```

This should then be followed by double loop similar to that used in the previous function to pair all combinations of operations and atoms in the input geometry, with new atoms that are discovered being added to the list:

```
atomnum_start = atomnum;
for k = 1:num_operations
  for j = 1:atomnum_start
    xyz_old = [x(j) y(j) z(j)]';
    xyz_new = operations(:,:,k)*xyz_old;
    foundit = 0;
    for m = 1:atomnum
      dist = norm(xyz_new - [x(m) y(m) z(m)]');
      if(dist < 0.001)
        foundit = 1;
      end
    end
    if(foundit == 0)
      atomnum = atomnum + 1;
      stop = 0;
      x(atomnum,1) = xyz_new(1,1);
      y(atomnum,1) = xyz_new(2,1);
      z(atomnum,1) = xyz_new(3,1);
      name(atomnum,:) = name(j,:);
    end
  end
end
```

Finally, the resulting list of atomic coordinates is written to the output file:

```
f2 = fopen(filename2,'w');

for j = 1:atomnum
  fprintf(f2,'%s%f%f%f\n',name(j,:),x(j,1),...
    y(j,1),z(j,1));
end

fclose(f2);
```

These steps are combined in the following function:

```
function apply_group(filename1, filename2, ...
operations)

num_operations = size(operations);
num_operations = num_operations(3);

[name,x,y,z] = textread(filename1,'%s %f %f %f');
name = char(name);
name
atomnum = size(x);
atomnum = atomnum(1);

atomnum_start = atomnum;
for k = 1:num_operations
  for j = 1:atomnum_start
    xyz_old = [x(j,1) y(j,1) z(j,1)]';
    xyz_new = operations(:,:,k)*xyz_old;
    foundit = 0;
    for m = 1:atomnum
      dist = norm(xyz_new - [x(m) y(m) z(m)]');
      if(dist < 0.001)
        foundit = 1;
      end
    end
    if(foundit == 0)
      atomnum = atomnum + 1;
      stop = 0;
      x(atomnum,1) = xyz_new(1,1);
      y(atomnum,1) = xyz_new(2,1);
      z(atomnum,1) = xyz_new(3,1);
      name(atomnum,:) = name(j,:);
    end
  end
end

f2 = fopen(filename2,'w');

for j = 1:atomnum
```

```
    fprintf(f2,'%s%f%f%f\n',name(j,:),x(j,1),...
       y(j,1),z(j,1));
 end

 fclose(f2);
```

Exercise 8.3. **Limitations on the combination of symmetry operations.**
As is clear from the while loop in the BuildPointGroup function,
the symmetry operations of a group should form a complete and closed
set. This requirement places restrictions on how different symmetry
operations can be combined in a point group. To see how things can go
wrong, let's try placing a $\hat{C}_3$ operation and a perpendicular $\hat{C}_4$ operation
together in a point group. (a) Set up as generating matrices a $\hat{C}_3$ rotation
around z, and a $\hat{C}_4$ rotation around x. (b) Create a modified version of the
BuildPointGroup function that will exit the while loop after four
cycles if the point group is not completed by then. (c) Run this modified
function using the $\hat{C}_3$ and $\hat{C}_4$ matrices as generators. How many operations result? (d) Use the apply_group function to apply these operations to a H atom at the point [1.0 0.75 0.50]. Plot the structure of the
resulting geometry, and describe what has happened.

BUILDING THE MOLECULAR SYMMETRY GROUPS

The requirement of the operations within a group forming a closed set
leads to there being only a limited number of types of molecular point
groups. These can be categorized by their generating operations. C_1 is the
group corresponding to trivial symmetry, with its only operation being
$\hat{E}$. The groups C_n, C_s, C_i, and S_n (n = even) add to $\hat{E}$ one addition generating operation: $\hat{C}_n$, $\hat{\sigma}$, $\hat{i}$, and $\hat{S}_n$, respectively. Further generating operations
can be added to the C_n point group which leave the high-symmetry axis
(usually chosen as z) unchanged. C_{nv} and C_{nh} add an $\hat{\sigma}$ operation whose
mirror plane lies parallel or perpendicular to the C_n-axis, respectively.
The D_n point groups, on the other hand, add to their corresponding C_n
point groups a $\hat{C}_2$ generating operation whose axis lies perpendicular to
the original C_n-axis. The D_n point groups can then be expanded up by
the addition of an $\hat{\sigma}$ operation whose mirror plane is perpendicular to
the C_n-axis to give the D_{nh} point groups, or whose mirror plane bisects a
pair of the perpendicular C_2-axes of the D_n group, yielding the D_{nd} point
groups.

Finally there is a limited number of high symmetry point groups in which multiple C_n-axes are present. These are based on the symmetries of the tetrahedron, octahedron, and icosahedron. The full symmetries of these polyhedra are given by T_d, O_h, and I_h, respectively, while their rotational subgroups (with no generating operation involving a reflection or inversion) are T, O, and I. One additional tetrahedral point group is created by adding a mirror operation to T, whose plane lies perpendicular to the one of the C_2-axes of the tetrahedron.

The previous two paragraphs were dense in information, but highlight that the range of point groups for a molecular system is limited. To translate these short descriptions into a full picture of the symmetry they imply, it is helpful to have a function which plots how the operations of a point group transform a single point. Such a function is given here:

```
function make_projection(start_position,...
  outputfile,operations)

num_operations = size(operations);
num_operations = num_operations(3);

atomnum = 1;
x(1,1) = start_position(1);
y(1,1) = start_position(2);
z(1,1) = start_position(3);

atomnum_start = atomnum;
stop = 0;
loop_counter = 0;
while (stop == 0)
  stop = 1;
  for k = 1:num_operations
    for j = 1:atomnum_start
      xyz_old = [x(j,1) y(j,1) z(j,1)]';
      xyz_new = operations(:,:,k)*xyz_old;
      foundit = 0;
      for m = 1:atomnum
        dist = norm(xyz_new - [x(m,1) y(m,1) z(m,1)]');
        if(dist < 0.001)
          foundit = 1;
        end
      end
```

```
      if(foundit == 0)
        atomnum = atomnum + 1;
        stop = 0;
        x(atomnum,1) = xyz_new(1,1);
        y(atomnum,1) = xyz_new(2,1);
        z(atomnum,1) = xyz_new(3,1);
      end
    end
  end
  atomnum_start = atomnum;
  loop_counter = loop_counter + 1;
  fprintf('Loop %d: total # of atoms = %d\n',...
    loop_counter, atomnum);
end

f2 = fopen(outputfile,'w');
for j = 1:atomnum
  if(z(j,1) >= 0)
    name_atom = 'C';
  end
  if(z(j,1) <= 0)
    name_atom = 'H';
  end
  fprintf(f2,'%s %f %f %f\n',name_atom,x(j,1),...
    y(j,1),z(j,1));
end
fclose(f2);
figure
axis equal;
axis off;
hold on;
draw_spheres(outputfile,'C',[0, 0, 0],0.1,40);
draw_spheres(outputfile,'H',[0.7 0.7 0.7],0.1,40);
% Plot circle in x-y plane to aide in
% visualization.
r = 2.0;

x_coord = -r:r/100:r;

y_coord = (r^2*ones(size(x_coord))-x_coord.*...
  x_coord).^0.5;
z_coord = 0*x_coord;
hold on
```

```
plot3 (x_coord,y_coord,z_coord,'color',[0,0,0],...
   'linewidth',2);
plot3 (x_coord,-y_coord,z_coord,'color',[0,0,0],...
   'linewidth',2);
plot3 (0,0,0,'.','color',[0,0,0],'MarkerSize',20);

% Set view slightly off of z axis.
view([.05,.05,1]); camup([0,1,0]);
light
```

Exercise 8.4. Annotate the above function to explain how it works and what it does.

Exercise 8.5. Use the make_projection function to show how the following point groups transform a general point, say [1.0 0.75 0.50]: C_4, C_{4v}, C_{4h}, D_4, D_{4h}, D_{4d}, C_5, C_{5v}, C_{5h}, D_5, D_{5h}, and D_{5d}. Annotate the plots with all symmetry elements that you can identify and give the orders of the groups.

CHAPTER 9

Group Theory and Basis Sets

INTRODUCTION

We have just seen how matrices describing the effect of symmetry operations on a position can be used to represent a point group. The power of group theory to simplify the analysis of chemical problems becomes clearer when we turn to how matrices can capture the tranformations of a basis set of functions under symmetry operations. In this chapter, we will begin to explore how the basis set used in describing the bonding in a molecule forms a representation of that molecule's point group symmetry, and how group theory shows us how to rearrange the basis set into subsets that only interact with each other. To illustrate this, we will focus our attention on a simple model system, the sp^3 hybrid orbitals of the oxygen atom in a water molecule. As we approach more complex molecules in future chapters, we will see how the principles described here are broadly generalizable.

sp^3 HYBRID ORBITALS OF H$_2$O AS A BASIS FOR REPRESENTING POINT GROUP SYMMETRY

In a localized bonding description of the H$_2$O molecule, the 2s and 2p orbitals of the oxygen atom are considered to combine to form four sp^3 hybrid orbitals that point toward the corners of a tetrahedron. Two of these hybrid orbitals point toward H atoms, and are used in the formation of O-H σ bonds. The remaining two hybrid orbitals point toward empty space, hosting lone pairs of electrons. Here, we will use these four O sp³ hybrid orbitals to illustrate the application of group theory to a basis set.

To start, we will need to generate the hybrid orbitals from the valence s and p orbitals of the O atom. A trick for doing this is to recognize that a tetrahedron can be easily generated from a cube of points, when every other corner of the cube is taken. Consider the following coordinates in which a cube is built from C and N atoms (as reference points, not part of the chemical system), with the two elements alternating on the corners:

```
N    1    1    1
C    1    1   -1
C    1   -1    1
C   -1    1    1
N    1   -1   -1
N   -1    1   -1
N   -1   -1    1
C   -1   -1   -1
```

This geometry can be plotted with the following MATLAB® code:

```
figure
draw_bonds('cube-geo','C',[0,0,0],'N',...
            [0 0 0],0.1,3.0,4);
axis off; axis equal
draw_spheres('cube-geo','C',[0,0,0],.1,40);
draw_spheres('cube-geo','N',[0,90,255]/255,.1,40);
light
view([1 -.5 -.2]);camup([0 0 -1]);
```

which should yield an image similar to that shown in Figure 9.1. In this figure, the C and N atoms appear as black and blue spheres, respectively. If you rotate this image around in the MATLAB Figure window, it will quickly become clear that the blue atoms define the corners of an ideal tetrahedron, as do the black atoms.

sp³ hybrid orbitals are created by taking linear combinations of the p_x, p_y, and p_z orbitals that point toward each of the corners for one of these tetrahedra, and giving an equal contribution of the s orbital to each of these linear combinations to create constructive interference for each hybrid

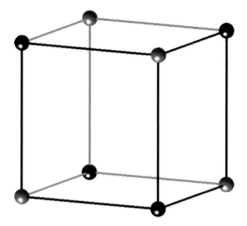

FIGURE 9.1   Cube defined by the coordinates in the text.

orbital in the direction of the chosen corner. For example, hybrid orbitals pointing toward the blue corners are defined as:

$$\psi_{sp^3,1} = \frac{1}{2}\left(\psi_{2s} + \psi_{2p_x} + \psi_{2p_y} + \psi_{2p_z}\right)$$

$$\psi_{sp^3,2} = \frac{1}{2}\left(\psi_{2s} - \psi_{2p_x} - \psi_{2p_y} + \psi_{2p_z}\right)$$

$$\psi_{sp^3,3} = \frac{1}{2}\left(\psi_{2s} + \psi_{2p_x} - \psi_{2p_y} - \psi_{2p_z}\right)$$

$$\psi_{sp^3,4} = \frac{1}{2}\left(\psi_{2s} - \psi_{2p_x} + \psi_{2p_y} - \psi_{2p_z}\right)$$

which in the full basis set for a H_2O molecule (O 2s, O $2p_x$, O $2p_y$, O $2p_z$, H 1s on first H atom, H 1s on second H atom) could be expressed in MATLAB as:

```
sp3_1 = 0.5*[1  1  1  1  0  0]';
sp3_2 = 0.5*[1 -1 -1  1  0  0]';
sp3_3 = 0.5*[1  1 -1 -1  0  0]';
sp3_4 = 0.5*[1 -1  1 -1  0  0]';
```

Exercise 9.1. Show that the four functions hybrid orbitals form an orthonormal set. *Hint*: This is most easily demonstrated by taking matrix dot products among the above vectors.

How would the water molecule fit in this picture? The O atom should be at the origin (center of the cube), with the H atoms placed near the ends of two of these hybrid orbitals. To speed things along, I will save you the trigonometric work involved in calculating the H atom coordinates, and give you the following to save in a file named H2O-geo:

```
O    0.0        0.0        0.0
H    0.5356     0.5356     0.5871
H   -0.5356    -0.5356     0.5871
```

This molecule can be plotted in the context of the cube from above using the following commands:

```
hold on
draw_spheres('H2O-geo','O',[1 0 0],.1,40)
draw_spheres('H2O-geo','H',[.8 .8 .8],.05,40)
draw_bonds('H2O-geo','O',[1 0 0],'H',[.8 .8 .8],...
           0.1,1.0,6);
light
axis equal
axis off
view([1 -.5 -0.2]);camup([0 0 -1]);
```

which yield the picture in Figure 9.2. This image will form the foundation for many other figures in this chapter. The process of reusing it repeatedly

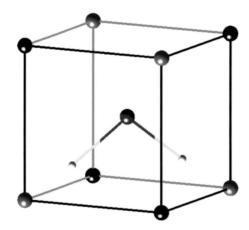

FIGURE 9.2 Water molecule oriented inside of a cube whose blue corners define the directions of the O atom's sp³ hybrid orbitals.

is made simpler by saving it as a `.fig` file using the menus of the Figure window (or the `save` function at the command line). Using the `open` command, a new Figure window can then be opened containing these geometrical elements.

With this figure template in hand, we are almost ready to start plotting hybrid orbitals. One additional step that makes this process more convenient is to make a few edits to our `drawMO` function so that: (1) the atomic coordinates are read from our geometry file, and (2) the new figure command is removed to allow the orbitals to be added to the existing figure that MATLAB is currently working with:

```
function drawMO(coordinates_file, orbital_counts,...
  orbital_coefficients, scale_factor)

[name,x,y,z] = textread(coordinates_file, ...
    '%s %f %f %f');

coordinates=[x y z];
figure
. . .
```

The four hybrid orbitals can then be visualized one-by-one by using sequences of commands such as the following:

```
>> open H2OinCube.fig
>> pause(1) % give time to figure window to be
>> % read before plotting.
>> drawMO('H2O-geo',[4 1 1], sp3_1, 1)
>> view([1 -.5 -.2]);camup([0 0 -1]);
>> title('\fontsize{18} lone pair 1');
```

Exercise 9.2. Use this process to plot the four sp^3 hybrid orbitals created above. You should obtain images quite similar those shown in Figure 9.3.

BASIS SETS AS REPRESENTATIONS OF POINT GROUP SYMMETRY

We are now going to use these four hybrid functions as a representation of the C_{2v} symmetry of a water molecule. When we restrict our basis set to these four hybrid orbitals, we can express a general function in terms

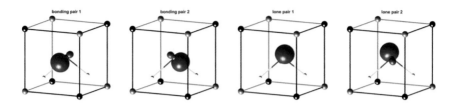

FIGURE 9.3   sp³ hybrid orbitals defined for the O atom of a water molecule.

of coefficients corresponding to the contribution from each hybrid orbital. The hybrids can then be written in MATLAB as basis vectors in a 4D space:

```
bp1 = [1 0 0 0]';
bp2 = [0 1 0 0]';
lp1 = [0 0 1 0]';
lp2 = [0 0 0 1]';
```

And more complicated linear combinations of these hybrid orbitals would then be expressed as other vectors in this space.

   Note that these functions are no longer written in terms of atomic orbitals, and will need to be translated back to the atomic orbital representation when we want to plot them or calculate Hamiltonian matrix elements between them. Here is an example of some code that would carry out this translation for plotting the functions:

```
function plot_in_basis(geofile,orblist,...
  psi_in_basis,basis,scale_factor)

num_basis_functions = size(basis);
num_basis_functions = num_basis_functions(2);
num_atomic_orbitals = sum(orblist);

psi_ao = zeros(num_atomic_orbitals,1);

for j = 1:num_basis_functions
  psi_ao = psi_ao + psi_in_basis(j,1)*basis(:,j);
end

drawMO(geofile,orblist,psi_ao,scale_factor);
```

which could be used in the following way:

```
open H2OinCube.fig
pause(1)
plot_in_basis('H2O-geo',[4 1 1], lp1, [sp3_1 ...
  sp3_2 sp3_3 sp3_4], 1)
view([1 -.5 -.2]);camup([0 0 -1]);
title('\fontsize{18} bonding pair 1');
```

to give plots of the hybrid orbitals as we made in Exercise 9.2.

Exercise 9.3. Annotate the code for the plot_in_basis function to explain the purpose of each line of code, and how it achieves the translation of a wavefunction represented as a linear combination of sp³ hybrid orbitals basis back into the atomic orbital representation.

We now have our basis set of sp³ hybrid orbitals ready. A group theoretical analysis of this basis begins by constructing matrices that show how the basis functions are transformed into each other by the symmetry operations of the point group. The full collection of these matrices comprise a **matrix representation** of the point group. In this specific case, the matrices would carry out the following operations:

$$\begin{pmatrix} c_{sp^3,1}' \\ c_{sp^3,2}' \\ c_{sp^3,3}' \\ c_{sp^3,4}' \end{pmatrix} = \begin{pmatrix} R_{11} & R_{12} & R_{13} & R_{14} \\ R_{21} & R_{22} & R_{23} & R_{24} \\ R_{31} & R_{32} & R_{33} & R_{34} \\ R_{41} & R_{42} & R_{43} & R_{44} \end{pmatrix} \begin{pmatrix} c_{sp^3,1} \\ c_{sp^3,2} \\ c_{sp^3,3} \\ c_{sp^3,4} \end{pmatrix}$$

Where R_{ij} shows how much of $\psi_{sp^3,j}$ ends up contributing to $\psi_{sp^3,i}$ in the transformed function.

Exercise 9.4. Using the plot_in_basis function, confirm that the following matrices properly represent the symmetry operations of the C_{2v} point group.

```
E_sp3 = [ 1 0 0 0
          0 1 0 0
          0 0 1 0
          0 0 0 1 ];
```

```
C2_sp3 = [ 0 1 0 0
           1 0 0 0
           0 0 0 1
           0 0 1 0];

sigma1_sp3 = [ 0 1 0 0
               1 0 0 0
               0 0 1 0
               0 0 0 1];

sigma2_sp3 = [ 1 0 0 0
               0 1 0 0
               0 0 0 1
               0 0 1 0];
```

CHARACTERS OF A MATRIX REPRESENTATION

Constructing these matrices to represent the symmetry operations of the C_{2v} point group was very manageable for two reasons: We only had four basis functions, and the point group only has four operations. In most cases, however, the situation is not as simple, and constructing such matrices would be a time-consuming process. Fortunately, most of the information we will need about these matrices is contained in the traces of the matrices, the sum of the elements of each matrix along the diagonal:

$$\chi(\hat{R}) = \sum_j R_{jj}$$

In the context of group theory, the trace of a matrix representing a symmetry operation is known as the **character** of the matrix. These characters can often be determined quickly without needing to build the full matrices. In the case of the sp³ hybrid orbitals, the $\hat{E}$ operation leaves all four functions as they are, so $\chi(\hat{E}) = 4$. The $\hat{C}_2$ operation interchanges the bonding pairs and lone pairs, so that no function is left where it started; $\chi(\hat{C}_2)$ is thus 0. The mirror operations interchange either the bonding pairs or the lone pairs, leaving the other functions unchanged, so $\chi(\hat{\sigma}_1) = \chi(\hat{\sigma}_2) = 2$. The full list of characters can be summarized in vector form:

$$\bar{\chi} = \left(\chi(\hat{E}) \quad \chi(\hat{C}_2) \quad \chi(\hat{\sigma}_1) \quad \chi(\hat{\sigma}_2) \right) = (4 \quad 0 \quad 2 \quad 2)$$

Exercise 9.5. Show using MATLAB's `trace` command (as well as count-ing in your head) that these characters agree with the traces of the matri-ces found in Exercise 9.4.

REDUCIBLE AND IRREDUCIBLE REPRESENTATIONS

In looking through the matrices we made for representing the symme-try operations that apply to our sp³ hybrid orbitals, you might notice that the lone pair functions and the bonding pair functions are never inter-changed. This is clearly seen in that matrices have zeros for all elements that involve mapping between these two types of functions:

$$\hat{E} = \begin{pmatrix} 1 & 0 & 0 & 0 \\ 0 & 1 & 0 & 0 \\ 0 & 0 & 1 & 0 \\ 0 & 0 & 0 & 1 \end{pmatrix}, \quad \hat{C}_2 = \begin{pmatrix} 0 & 1 & 0 & 0 \\ 1 & 0 & 0 & 0 \\ 0 & 0 & 0 & 1 \\ 0 & 0 & 1 & 0 \end{pmatrix},$$

$$\hat{\sigma}_1 = \begin{pmatrix} 0 & 1 & 0 & 0 \\ 1 & 0 & 0 & 0 \\ 0 & 0 & 1 & 0 \\ 0 & 0 & 0 & 1 \end{pmatrix}, \quad \hat{\sigma}_2 = \begin{pmatrix} 1 & 0 & 0 & 0 \\ 0 & 1 & 0 & 0 \\ 0 & 0 & 0 & 1 \\ 0 & 0 & 1 & 0 \end{pmatrix}$$

The matrices are built from smaller 2×2 blocks that never interact with each other, a property which is referred to by the term **block diagonal**. These sets of smaller 2×2 matrices on their own could be used as two separate representations of C_{2v} symmetry. Operations on the sp³ hybrid basis set can be broken down into smaller matrices that also represent the C_{2v} symmetry. These 4×4 matrices provide an example of a **reduc-ible representation**. Representations are reducible when they can be block diagonalized through a similarity transform ($\hat{R}' = \hat{A}^{-1}\hat{R}\hat{A}$, where the same $\hat{A}^{-1}...\hat{A}$ pair is applied to all operations in the group) to give a series of simpler representations.

What do these similarity transforms accomplish? Recall that similarity transforms are the tools we used before to make symmetry operations that are oriented along directions other that the principal axes of the coordi-nate system. Similarity transforms, then, correspond to the reorientation of an operation within a vector space. In this case, however, the vector space represents the wavefunctions that can be built from our basis set, rather than points in Cartesian space.

We can get a better sense of what is happening here by applying a simple trick to the equation showing how a wavefunction is affected by a symmetry operation: $\psi' = \hat{R}\psi$. In this equation we can insert $\hat{A}\,\hat{A}^{-1}$ whereever we like, as its net effect is an identity. So, $\psi' = \hat{R}\psi = \hat{R}\,\hat{A}\,\hat{A}^{-1}\psi$. If we then apply $\hat{A}^{-1}$ to the left-hand side of both sides of the equation, we obtain:

$$\hat{A}^{-1}\psi' = \hat{A}^{-1}\hat{R}\,\hat{A}\,\hat{A}^{-1}\psi$$

which can be written as:

$$\left(\hat{A}^{-1}\psi'\right) = \left(\hat{A}^{-1}\hat{R}\,\hat{A}\right)\left(\hat{A}^{-1}\psi\right) = \hat{R}'\left(\hat{A}^{-1}\psi\right)$$

Here, we see that the new operations created by the similarity transformation are symmetry operations for a new basis set obtained by taking linear combinations of the original basis set functions. The formula for taking these linear combinations is encoded in $\hat{A}^{-1}$.

The thought of determining an $\hat{A}^{-1}$ matrix with these properties may sound daunting. But the process of breaking down a reducible representation is made simpler by the fact that the characters of the matrices (which are unaffected by a similarity transform) are all we need for this process. In the next sections, we will see how the characters of a representation can be used to decompose a representation into a linear combination of **irreducible representations**, the representations of each point group that cannot be further broken down.

REDUCTION OF REDUCIBLE REPRESENTATIONS

In the last section, we saw that the process of reducing a reducible representation involves finding a set of linear combinations of the basis functions such that the matrices describing their interchange under symmetry operations become block diagonal. The end result of this process is that the reducible representation is resolved into a sum of irreducible representations. The key to the procedure for doing this is to consider the characters for the reducible representation as a vector in an abstract space ($\bar{\chi}_{\text{red}}$).

The vectors of the irreducible representations have been cataloged in character tables, which are available from many sources (for example, the website www.webqc.org/symmetry.php). The table for the C_{2v} point group is given here:

C_{2v}	$\hat{E}$	$\hat{C}_2$	$\hat{\sigma}_1$	$\hat{\sigma}_2$
A_1	1	1	1	1
A_2	1	1	−1	−1
B_1	1	−1	1	−1
B_2	1	−1	−1	1

As you can confirm from this table, the character vectors for the irreducible representations, $\bar{\chi}_{irred,i}$, satisfy relationships $\bar{\chi}_{irred,i} \cdot \bar{\chi}_{irred,j} = h\,\delta_{ij}$, where h is the order of the group. In this way, the $\bar{\chi}_{irred,i}$ vectors form an orthogonal set of vectors, which serve as basis vectors for this abstract space. Any vector in this space can then be written as linear combinations of the $\bar{\chi}_{irred,i}$ set.

What is the dimensionality of this space? It will be equal to the number of coefficients in the vectors that can be independently varied. In some cases, this will just be the order of the group, as in C_{2v}. However, in higher symmetry cases, symmetry relationships between different operations can make their characters equal: Symmetry operations of the same class are related by similarity transforms, which leave the characters of the matrices invariant. There will only be one independently varying character for each class in a point group. In general, the dimensionality will then be equal to the number of classes in the point group. The number of irreducible representations necessary to map the representation space will then be the number of classes.

With the irreducible representations being tabulated, we can get the contribution of each irreducible representation to a reducible one by just projecting the reducible representation vector onto the irreducible ones. The reducible representation is considered to be a linear combination of irreducible ones, so that its character vector can be written as $\bar{\chi}_{red} = \sum_i a_i \bar{\chi}_{irred,i}$. The coefficients in this expansion are then obtained as:

$$a_j = \frac{\left(\bar{\chi}_{irred,j} \cdot \bar{\chi}_{red}\right)}{h} = \bar{\chi}_{irred,j} \cdot \frac{\left(\sum_i a_i \bar{\chi}_{irred,i}\right)}{h} = \sum_i a_i\, \bar{\chi}_{irred,i} \cdot \frac{\bar{\chi}_{irred,j}}{h} = \sum_i \frac{a_i h \delta_{ij}}{h} = a_j$$

This process can be easily implemented in MATLAB to make a function that will give the irreducible components of any reducible representation:

```
function reduce_representation(characters,...
    character_table_byclass,class_mult,IRnames)

group_order = sum(class_mult);
num_classes = size(character_table_byclass);
num_classes = num_classes(2);
contribution = zeros(num_classes,1);
for j = 1:num_classes
  contribution(j,1) = characters*...
(character_table_ byclass(j,:).*class_mult)'/...
group_order;
end

fprintf('\n');
fprintf('Group order = %d\n',group_order);
fprintf('Reducible representation = ');
for j = 1:num_classes
  if(contribution(j,1) ~= 0)
    fprintf(' + %d*%s',contribution(j,1),...
      IRnames(j,:));
  end
end
fprintf('\n\n');
```

Exercise 9.6. Annotate the reduce_representation function to explain how it achieves the decomposition described above. Explain what formats the input variables are given in (e.g. row vector, column vector, matrix) and how the information should be distributed in these variables.

Exercise 9.7. Confirm that the output of the function for the sp³ hybrid orbital basis in water is the following:

```
>> reduce_representation(x_sp3_vect,...
C2v_characters,C2v_mult,C2v_names)

Group order = 4
Reducible representation = + 2*a1 + 1*b1 + 1*b2
```

In future chapters, we will apply this program to more complicated examples.

TRANSFORMATION OF BASIS SET TO IRREDUCIBLE REPRESENTATIONS

Once the irreducible components to a reducible presentation are determined, the character table provides instructions for taking the proper linear combinations of the original basis functions to create those irreducible representations. For example, the A_1 representation of C_{2v} symmetry is symmetric with respect to all symmetry operations. The two A_1 components of the sp³ hybrid orbital basis set should then be of the forms $\psi_{lp1} + \psi_{lp2}$ and $\psi_{bp1} + \psi_{bp2}$. The B_1 and B_2 components are then built from the corresponding differences.

This process of using the characters of the irreducible representation as a guide to taking linear combinations of the basis functions that transform according to that representation (symmetry adapted linear combinations, SALCs) is formalized in the **projection operator** method:

$$\psi_{SALC,\,irred\,i} = \left(\sum_{\hat{R}} \chi^{*}_{irred\,i}(\hat{R})\hat{R} \right) \psi_{starting\,point}$$

When several functions of the same irreducible representation are present, they can be generated by using different basis functions as starting points.

Exercise 9.8. Annotate the following function, and use it to create SALCs for the sp³ hybrid orbital basis set of the O atom in H_2O. Use the plot_in_basis function to make plots of the SALCs.

```
function psi_SALC = projection_operator...
  (startpoint,matrix_representation, IR_characters)

num_operations = size(IR_characters);
num_operations = num_operations(2);
num_basis_functions = size(startpoint);
num_basis_functions = num_basis_functions(1);

psi_SALC = zeros(num_basis_functions,1);

for j = 1:num_operations
  psi_SALC = psi_SALC + IR_characters(1,j)...
    *matrix_representation(:,:,j)*startpoint;
end

% Normalize final wave_function

psi_SALC = psi_SALC/(psi_SALC'*psi_SALC)^0.5;
```

CHAPTER 10

The MOs of H$_2$O

INTRODUCTION

In the last chapter, we saw that the sp^3 hybrid orbitals on the O atom of an H$_2$O molecule form a reducible representation of the molecule's C_{2v} point group symmetry. Through calculating projections of this representation onto the irreducible representations of the point group, we found that the sp^3 hybrid basis set can be expressed as linear combinations of two a_1 functions, one b_1 function, and one b_2 function. The projection operator method then allowed us to construct these functions in Exercise 9.8, which should have yielded the results shown in Figure 10.1. With this result, and the basic outline of group theory that we just built, we are now ready to see how insights from symmetry can be implemented in electronic structure analysis.

A major way that symmetry relates to quantum mechanics can be seen by applying one of a molecule's symmetry operations to the Schrödinger equation specifying its stationary electronic states. We begin with $\hat{H}\psi_j = E_j\psi_j$, and then apply the symmetry transformation $\hat{R}$ to both sides:

$$\hat{R}\left(\hat{H}\psi_j\right) = \hat{R}\left(E_j\psi_j\right) = E_j\hat{R}\left(\psi_j\right)$$

Now, the Hamiltonian operator represents the physical situation in which the electrons move. If $\hat{R}$ is indeed a symmetry operation, this physical situation will be unchanged when the operation is applied. We thus obtain:

$$\hat{R}\left(\hat{H}\psi_j\right) = \hat{H}\left(\hat{R}\psi_j\right) = E_j\left(\hat{R}\psi_j\right)$$

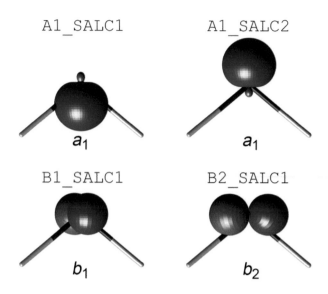

FIGURE 10.1 Symmetry adapted linear combinations (SALCs) constructed from sp³ hybrid orbitals on the O atom of an H_2O molecule.

or $\hat{H}\psi_j' = E_j\psi_j'$ with $\psi_j' = \hat{R}\psi_j$. In other words, if ψ_j is an eigenfunction of the Hamiltonian operation, then applying symmetry operations to it will give rise to functions that are also eigenfunctions with the same eigenvalue.

This conclusion places strict conditions on the form of ψ_j. In the case where ψ_j is a singly degenerate function, there are no other functions that have the same eigenvalue. This means that $\left|\psi_j'\right|^2 = \left|\psi_j\right|^2$, which for a two-fold operation would imply that $\hat{R}\psi_j = \pm\psi_j$. If you were to write characters for these operations on ψ_j, you would get a string of +1's and −1's, just like those of the irreducible representations of the C_{2v} point group. In fact, *the eigenfunctions of the Hamiltonian operator in general transform as irreducible representations of the molecule's symmetry.* Combining wavefunctions that transform as different irreducible representations would create a function that no longer transforms as an irreducible representation, violating the above principle. As such, *two functions transforming as different irreducible representations of a system's symmetry have no interactions with each other:* The overlap integrals and Hamiltonian matrix elements for their interactions will always be zero.

In this chapter, we will see how these ideas can be applied with the construction of the MO diagram for H_2O. First, we will set up the Hamiltonian matrix in the atomic orbital basis, and quickly obtain the MOs to provide a reference point for our future work. Then, we will combine the SALCs of our O sp^3 hybrid orbitals from the last chapter with SALCs for the H 1s orbitals to see how H_2O's MO diagram emerges from a series of orbital interactions.

THE MOs OF H₂O BY BRUTE FORCE

Before we commence with the application of symmetry to the H_2O molecule, let's take a quick look at the answers we should aim for. The Hamiltonian matrix for the molecule can be obtained using the `build_hamiltonian` function we developed previously, and the MOs and their energies can then be calculated with `eig`. For this calculation, you may use the following Hückel parameters that have been specifically calibrated against the results of a DFT calculation on crystalline H_2O:

```
if(strcmp(atomname(j),'H')==1)
 params(j,:) = [1 -6.692 2.5981 0 0.0 0.0];
 num_orbitals = num_orbitals+1;
 orb_list(1,j) = 1;
 foundit = 1;
end
if(strcmp(atomname(j),'O')==1)
 params(j,:) = [2 -23.487 2.5055 2 -9.177 1.8204];
 num_orbitals = num_orbitals+4;
 orb_list(1,j) = 4;
 foundit = 1;
end
```

To confirm that your parameters and `build_hamiltonian` function are behaving properly, check your Hamiltonian and overlap matrices against those given here:

```
H_H2O =
```

-23.4870	0	0	0	-5.8575	-5.8575
0	-9.1770	0	0	-3.1063	3.1063
0	0	-9.1770	0	-3.1067	3.1067
0	0	0	-9.1770	-3.4050	-3.4050
-5.8575	-3.1063	-3.1067	-3.4050	-6.6920	-0.1853
-5.8575	3.1063	3.1067	-3.4050	-0.1853	-6.6920

```
S_H2O =

    1.0000        0          0          0     0.2218    0.2218
         0   1.0000          0          0     0.2237   -0.2237
         0        0     1.0000          0     0.2237   -0.2237
         0        0          0     1.0000     0.2452    0.2452
    0.2218   0.2237     0.2237     0.2452     1.0000    0.0158
    0.2218  -0.2237    -0.2237     0.2452     0.0158    1.0000
```

Exercise 10.1. Now that we have the Hamiltonian matrix, the MOs can be obtained by diagonalizing it. Construct the MO diagram for the H_2O molecule, and assign each of the wavefunctions to an irreducible representation of C_{2v} symmetry. You should be able to reproduce the diagram in Figure 10.2.

Where do these MOs come from? In the following, we will see how group theory can be used to gradually build this diagram through pairwise interactions between orbitals, each with their own clear chemical interpretation. The MO diagram above will serve as our answer key for this process, much like the picture on the cover of the box for a jigsaw puzzle.

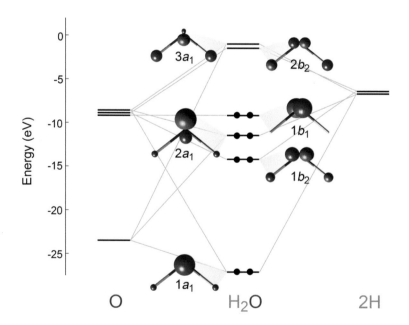

FIGURE 10.2 Simple Hückel molecule orbital diagram calculated for an H_2O molecule.

THE MOs OF H$_2$O FROM SP³ HYBRID SYMMETRY ADAPTED LINEAR COMBINATIONS (SALCs)

The SALCs of the sp³ hybrids we generated in the last chapter will be part of our basis set for building the MO diagram of H$_2$O. Where we last left them, they were encoded in 4×1 column vectors in MATLAB® (A1_SALC1, A1_SALC2, B1_SALC1, B2_SALC1), with each coefficient in the vectors giving the contribution from one of the original sp³ hybrid orbitals. As we start to apply these functions to the calculation of interactions, we will run into a compatibility issue. Our Hamiltonian matrix for the H$_2$O molecule refers to an atomic orbital basis instead of an sp³ one. We will thus need to convert the functions back to the atomic orbital basis. This procedure of converting a specialized basis set back to the standard atomic orbital one is common enough that it is worth writing a function to carry it out in an automated fashion:

```
function psi_ao = basis2ao(psi_in_basis,basis)

num_basis_functions = size(basis);
num_basis_functions = num_basis_functions(2);
num_atomic_orbitals = size(basis);
num_atomic_orbitals = num_atomic_orbitals(1);

psi_ao = zeros(num_atomic_orbitals,1);

for j = 1:num_basis_functions
  psi_ao = psi_ao + psi_in_basis(j,1)*basis(:,j);
end
```

Exercise 10.2. Note that the above function takes a single function in a given basis set and translates it into a vector referring to an atomic orbital basis set. Write a generalized version of this function that can perform the same process on a row of column vectors, allowing you to convert an array of wavefunctions into an atomic orbital basis with a single command.

With the basis2ao function, we can then easily convert these SALC functions to an atomic orbital representation:

```
>> A1_SALC1_ao = basis2ao(A1_SALC1,[sp3_1 sp3_2 ...
    sp3_3 sp3_4]);
>> A1_SALC2_ao = basis2ao(A1_SALC2,[sp3_1 sp3_2 ...
    sp3_3 sp3_4]);
```

```
>> B1_SALC1_ao = basis2ao(B1_SALC1,[sp3_1 sp3_2 ...
   sp3_3 sp3_4]);
>> B2_SALC1_ao = basis2ao(B2_SALC1,[sp3_1 sp3_2 ...
   sp3_3 sp3_4]);
```

Our basis functions for the O atom are now ready. Next, we simply need to set up SALCs for the H atoms, which will be the goal of the next exercise.

Exercise 10.3. Consider a basis consisting of H 1s orbitals on each of the H atoms of the H_2O molecule. Reduce this basis to its irreducible components using the reduce_representation and projection_operator functions from the last chapter. For the former, you should determine the characters for the 2×2 matrices giving the transformations of these H 1s orbitals under the symmetry operations of the molecule, while for the latter you will need the matrices themselves. Show that the resulting SALCs are those given in Figure 10.3.

Feel free to adjust your functions by a factor of -1 to obtain the same relative phases between the H- and O-centered SALCs as those depicted in the above figures.

We now have the full basis set consisting of functions A1_SALC1_ao, A1_SALC2_ao, A1_SALC3_ao, B1_SALC1_ao, B2_SALC1_ao, and B2_SALC2_ao. For the remainder of this section, we will follow a series of exercises that work with this basis set to gradually approach a diagonalized Hamiltonian matrix and the final MOs for H_2O.

Exercise 10.4. Create Hamiltonian and Overlap matrices for the H_2O molecule in the SALC basis that we have generated. Show that the matrix elements between the various SALCs are zero unless the two functions transform as the same irreducible representation of the molecule's C_{2v} symmetry.

A1_SALC3_ao   B2_SALC2_ao

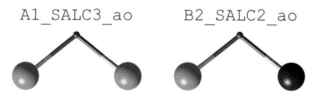

FIGURE 10.3   SALCs generated from the H 1s orbitals of an H_2O molecule.

Exercise 10.5. Build separate Hamiltonian matrices for the interactions among the functions of the same irreducible representations. You should obtain the following:

```
H_A1 =

-16.3320   -7.1550   -9.2624
 -7.1550  -16.3320   -2.4525
 -9.2624   -2.4525   -6.8773

H_B1 =

 -9.1770

H_B2 =

 -9.1770    6.2130
  6.2130   -6.5067
```

Exercise 10.6. Show that the interactions within the Hamiltonian for the b_2 SALCs lead to functions in Figure 10.4. Comment on the relative contributions of O 2p and H 1s functions to the lower and higher energy functions.

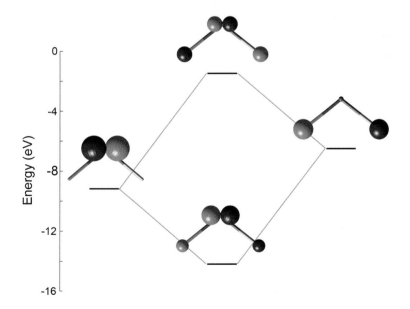

FIGURE 10.4   Two H_2O MOs derived from the interaction between the b_2 SALCS.

Hint: The MOs resulting from the diagonalization of the b_2 Hamiltonian matrix can be easily plotted using commands of the form:

```
drawMO('H2O-geo', [4 1 1], basis2ao(psi_B2(:,1),...
   [B2_SALC1_ao B2_SALC2_ao]),1);
view([1,-1,0]); camup([0 0 -1])
```

Let's now turn to the a_1 set of functions. There are three basis functions of a_1 symmetry, with their interactions being given by a 3×3 Hamiltonian matrix:

```
H_A1 =

-16.3320   -7.1550   -9.2624
 -7.1550  -16.3320   -2.4525
 -9.2624   -2.4525   -6.8773
```

The solutions here could be obtained by diagonalizing this matrix using the eig function. We can gain more insight into the bonding of this molecule, however, by working through this Hamiltonian matrix in a step-wise manner. Note that the largest matrix element is found between the first and third a_1 basis function. Working through this interaction will take us then a long way toward the final wavefunctions.

To carry out the interaction between A1_SALC1_ao and A1_SALC3_ao, we first create a smaller Hamiltonian (representing a subspace for the full range of functions in the basis set) involving just matrix elements between these two functions:

```
>> H_A1_step1_mini(1,1) = ...
A1_SALC1_ao'*H_H2O*A1_SALC1_ao;
>> H_A1_step1_mini(1,2) = ...
A1_SALC1_ao'*H_H2O*A1_SALC3_ao;
>> H_A1_step1_mini(2,1) = ...
A1_SALC3_ao'*H_H2O*A1_SALC1_ao;
>> H_A1_step1_mini(2,2) = ...
A1_SALC3_ao'*H_H2O*A1_SALC3_ao;
>> H_A1_step1_mini

H_A1_step1_mini =

-16.3320   -9.2624
 -9.2624   -6.8773
```

The results from turning on this pair-wise interaction are then obtained with eig:

```
>> [A1_step1_psi, A1_step1_E] = ...
eig(H_A1_step1_mini)

A1_step1_psi =

 -0.8528   0.5222
 -0.5222  -0.8528

A1_step1_E =

 -22.0037         0
        0   -1.2055
```

The resulting wavefunctions are then converted back to the atomic orbital basis:

```
>> A1_step1_psi1_ao = basis2ao(A1_step1_psi(:,1),...
     [A1_SALC1_ao A1_SALC3_ao]);
>> A1_step1_psi2_ao = basis2ao(A1_step1_psi(:,2),...
     [A1_SALC1_ao A1_SALC3_ao]);
>> A1_step1_psi3_ao = A1_SALC2_ao;
```

where the last command carries over the unused A1_SALC2_ao function to our current set of a_1 functions.

Exercise 10.7. Carry out the above procedure for the interaction between A1_SALC1_ao and A1_SALC3_ao, and show that it can be represented with the diagram in Figure 10.5.

The interaction diagram above has a clear interpretation: The symmetric combination of H 1s orbitals interacts to form in-phase and out-of-phase combinations with the O s-p hybrid function that is directed toward them. That this interaction has the largest Hamiltonian matrix element for the a_1 SALC Hamiltonian makes sense from the strong overlap between these functions. Another notable feature here is that the bonding combination is rich in O character, while the antibonding combination exhibits stronger contributions from the H orbitals. This distribution is in line with the O hybrid orbital being lower in energy than the H 1s combination.

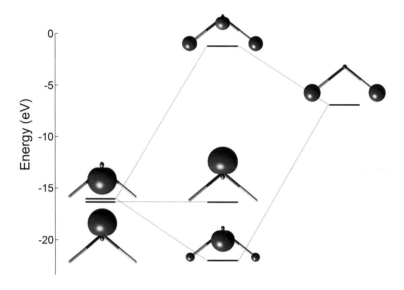

FIGURE 10.5   Step 1 in the process for deriving MOs from the a_1 SALCs of H_2O.

A comparison of these functions to the a_1 MOs in the completed MO diagram of water shows that we have made a lot of progress: We have already arrived at the overall form of the three-level bonding, non-bonding, antibonding combination found in the full MO diagram. The qualitative features are also similar, though our current bonding orbital has much less O 2s character than the final one, and we do not yet see any H-O interactions in the central non-bonding function.

We can check more quantitatively on what interactions are still unresolved by calculating the Hamiltonian matrix for our current a_1 functions. This matrix can be generated with the following commands (or with a single command using more sophisticated linear algebra):

```
>> H_A1_step2(1,1) = A1_step1_psi1_ao'*H_H2O...
   *A1_step1_psi1_ao;
>> H_A1_step2(1,2) = A1_step1_psi1_ao'*H_H2O...
   *A1_step1_psi2_ao;
>> H_A1_step2(1,3) = A1_step1_psi1_ao'*H_H2O...
   *A1_step1_psi3_ao;
>> H_A1_step2(2,1) = A1_step1_psi2_ao'*H_H2O...
   *A1_step1_psi1_ao;
>> H_A1_step2(2,2) = A1_step1_psi2_ao'*H_H2O...
   *A1_step1_psi2_ao;
```

```
>> H_A1_step2(2,2) = A1_step1_psi2_ao'*H_H2O...
   *A1_step1_psi3_ao;
>> H_A1_step2(2,2) = A1_step1_psi2_ao'*H_H2O...
   *A1_step1_psi2_ao;
>> H_A1_step2(2,3) = A1_step1_psi2_ao'*H_H2O...
   *A1_step1_psi3_ao;
>> H_A1_step2(3,1) = A1_step1_psi3_ao'*H_H2O...
   *A1_step1_psi1_ao;
>> H_A1_step2(3,2) = A1_step1_psi3_ao'*H_H2O...
   *A1_step1_psi2_ao;
>> H_A1_step2(3,3) = A1_step1_psi3_ao'*H_H2O...
   *A1_step1_psi3_ao;
>> H_A1_step2

H_A1_step2 =

   -22.0037     0.0000     7.3826
     0.0000    -1.2055    -1.6448
     7.3826    -1.6448   -16.3320
```

For the most part, this matrix has much less happening off the diagonal than in the previous step. The matrix elements with values of ca. −9 eV have become replaced with zeros. Now, the largest interaction remaining is that between the two lowest energy functions (the O-H bonding function, and the O-centered non-bonding function), with a matrix element of 7.38 eV.

To resolve this interaction, we then follow a similar sequence of commands as above. We first build a sub-space Hamiltonian:

```
>> H_A1_step2_MINI(1,1) = A1_step1_psi1_ao'*...
   H_H2O*A1_step1_psi1_ao;
>> H_A1_step2_MINI(1,2) = A1_step1_psi1_ao'*...
   H_H2O*A1_step1_psi3_ao;
>> H_A1_step2_MINI(2,1) = A1_step1_psi3_ao'*...
   H_H2O*A1_step1_psi1_ao;
>> H_A1_step2_MINI(2,2) = A1_step1_psi3_ao'*...
   H_H2O*A1_step1_psi3_ao;
```

Next, we diagonalize it, and convert the eigenvectors back to the atomic orbital basis:

```
>> [A1_step2_psi, A1_step2_E] = eig(H_A1_step2_MINI);
>> A1_step2_psi1_ao=basis2ao(A1_step2_psi(:,1),...
   [A1_step1_psi1_ao A1_step1_psi3_ao]);
```

```
>> A1_step2_psi2_ao=basis2ao(A1_step2_psi(:,2),...
    [A1_step1_psi1_ao A1_step1_psi3_ao]);
>> A1_step2_psi3_ao=A1_step1_psi2_ao;
```

The overall process can then be visualized with Figure 10.6.

Exercise 10.8. Reproduce the MO diagram in Figure 10.6, and explain what has been accomplished in this interaction that has led to the lower energy function being as low in energy as possible, and what has led to the nonbonding function going up in energy (whereas superficially, it looks as though it has acquired some O-H bonding).

With this interaction complete, let's check in again on the status of our Hamiltonian matrix. You should find that the updated Hamiltonian is:

```
H_A1_step3 =

  -27.0764          0   -0.9315
    0.0000   -11.2593    1.3557
   -0.9315     1.3557   -1.2055
```

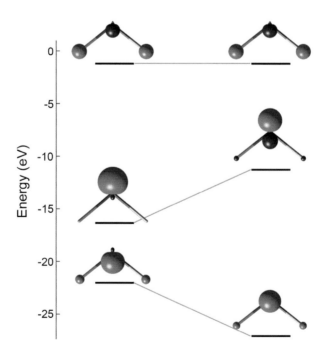

FIGURE 10.6 The second step in deriving MOs from the a_1 SALCs of an H_2O molecule.

where no off-diagonal elements significantly larger than 1 eV in magnitude are present. The weakness of the remaining interactions is evident in a comparison of our current a_1 functions with the true MOs for H_2O at the beginning of this chapter. Only minor differences remain.

Exercise 10.9. Another way to see that we have essentially converged on the correct MOs for H_2O is to go through one more round of diagonalizing the strongest remaining interaction in the Hamiltonian matrix. Perform this process, and determine how much the MO energies and shapes change as a result.

Overall, then, we see that only two interactions among our original a_1 SALCs are necessary to produce the nearly correct MOs for the H_2O molecule. This first one we need to turn on is the expected interaction between the H 1s orbitals and the O hybrid orbital directed toward them. The second interaction is less obvious: The lowest energy O sp hybrid orbital as generated contains an equal mixture of O 2s and 2p (you can confirm this by looking at the atomic orbital coefficients for the `A1_SALC1_ao` function), whereas the lowest energy function in the system would benefit from having a heavier contribution from the O 2s. This shift is accomplished by the interaction between the O-H bonding combination and the O sp non-bonding level.

PERCEIVING LOCALIZED BONDING IN H_2O

Overall, the use of sp³ hybrid functions on the O atom (when combined with group theory) has then provided a rather direct route to the MOs of the H_2O. We might wonder, then, to what degree we can consider the electronic structure of an H_2O molecule to be based on this hybrid orbital view. We'll explore this question in the next two exercises.

Exercise 10.10. Confirm that the linear combinations shown in Figure 10.7 give rise to localized bonding and non-bonding electron pairs. Compare the total energy of a system (in this approximation, the sum of the orbital energies weighted by their occupations) in which these four localized functions are occupied by electrons to that of an H_2O molecule with its first four MOs occupied.

Exercise 10.11. Build up a set of four functions corresponding to the hybrid orbital depiction of H_2O: Two sp³ hybrid lone pairs and two bonding functions created by combining an sp³ hybrid orbital with a H 1s orbital with

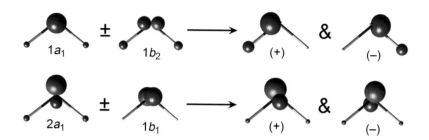

FIGURE 10.7 The generation of localized bonding and lone pair functions from the occupied MOs of H_2O.

equal coefficients ($\frac{1}{\sqrt{2}}$ and $\frac{1}{\sqrt{2}}$). Compare the energies of these functions with those of the localized functions obtained in the previous exercise.

BONUS CODE: BETTER BALL-AND-STICK MODELS

You might have noticed that in the images of this chapter, the bonds in the H_2O molecules have been plotted with 3D-rendered cylinders, rather than simple lines. Most of the tools necessary for making such plots became available to you when we discussed matrix transformations, as we will now see.

In MATLAB, a cylinder can be generated with the command:

```
[cylx cyly cylz] = cylinder();
```

where the `cylx`, `cyly`, and `cylz` variables give the x, y, and z coordinates, respectively, for a cylinder lying along the z-axis whose radius and length are both 1. To use such a function to build bonds, all we need to do is modify the position, orientation, and length of the cylinders that result from this function.

In our original `draw_bonds` function, we divided bonds into two vectors: One connecting the first atom to the midpoint of the point, and the other from the midpoint to the second atom. Let's focus just on the first vector for now. To adjust the length of the cylinder, we just multiply the z-coordinates by the length of the vector:

```
segment_vector = [midpoint(1)-x(j) midpoint(2)-...
    y(j)  midpoint(3)-z(j)]';
[cylx cyly cylz] = cylinder();
cylz = cylz*norm(segment_vector);
```

The width of the cylinder can be adjusted in a similar way:

```
cylx = cylx*width;
cyly = cyly*width;
```

Next, we can convert the vector to spherical polar coordinates to obtain the ϕ and θ angles for the orientation of the cylinder:

```
[phi, theta, r] = cart2sph(segment_vector(1,1),...
   segment_vector(2,1),segment_vector(3,1));
theta = pi/2-theta;
```

Rotations around the y and z axes can then be applied to all points in the cylinder to adjust the cylinder's axis to these values of ϕ and θ:

```
rotation1 = [cos(theta)      0     sin(theta)
                   0          1     0
             -sin(theta)      0     cos(theta) ];

rotation2 = [ cos(phi)   -sin(phi)      0
              sin(phi)    cos(phi)      0
                 0           0          1 ];

fullrotation = rotation2*rotation1;

npoints = size(cylx);
for j1 = 1:npoints(1)
  for k1 = 1:npoints(2)
r_old = [cylx(j1,k1) cyly(j1,k1) cylz(j1,k1)]';
r_new = fullrotation*r_old;
cylx_new(j1,k1) = r_new(1,1);
cyly_new(j1,k1) = r_new(2,1);
cylz_new(j1,k1) = r_new(3,1);
  end
end
```

Our cylinder is now of the correct length and the right orientation. We then simply need to move it to the correct location, and plot it as a surface:

```
surf(cylx_new+x(j),cyly_new+y(j), cylz_new+z(j),...
   'facecolor',color1,'edgecolor','none');
% Draw second half of bond with color2
```

The second segment of the bond is simply created by making another instance of this surface shifted to begin at the bond's midpoint rather than at atom 1, with a different color:

```
surf(cylx_new+x(j)+segment_vector(1,1),...
   cyly_new+ y(j)+segment_vector(2,1),...
   cylz_new+z(j)+segment_vector(3,1), ...
   'facecolor',color2,'edgecolor','none');
```

This code is implemented in the following modified version of the draw _ bonds function:

```
function draw_bonds_cylinder(filename,atom1,...
   color1,atom2,color2,d_min,d_max,width)

hold on
[atomname,x,y,z] = textread(filename,'%s %f %f %f');

natoms = size(x);
natoms = natoms(1);

atom_pos = [x y z];

for j = 1:natoms
  for k = 1:natoms
    if(strcmp(atom1,atomname(j))==1)
      if(strcmp(atom2,atomname(k))==1)
        dist = ( (x(j)-x(k))^2 + (y(j)-y(k))^2 + ...
                  (z(j)-z(k))^2 )^0.5;
      if( (dist<=d_max) && (dist>=d_min) )
        % TIME TO DRAW A BOND
        hold on;
        midpoint = [(x(j)+x(k))/2 (y(j)+y(k))/2...
                    (z(j)+z(k))/2];
        % Draw first half of bond with color1
        segment_vector = [midpoint(1)-x(j) ...
        midpoint(2)-y(j)  midpoint(3)-z(j)]';
        [cylx cyly cylz] = cylinder();
        cylx = cylx*width;
        cyly = cyly*width;
        cylz = cylz*norm(segment_vector);
        [phi, theta, r] = cart2sph(segment_vector(1,1),...
        segment_vector(2,1),segment_vector(3,1));
        theta = pi/2-theta;
```

```
% rotate by theta about y axis
rotation1 = [cos(theta) 0      sin(theta)
                  0      1      0
          -sin(theta)    0      os(theta) ];

rotation2 = [cos(phi)-sin(phi)0
               sin(phi)cos(phi)0
           0           0        1 ];

fullrotation = rotation2*rotation1;

npoints = size(cylx);
for j1 = 1:npoints(1)
    for k1 = 1:npoints(2)
        r_old = [cylx(j1,k1) cyly(j1,k1) ...
                cylz(j1,k1)]';
        r_new = fullrotation*r_old;
        cylx_new(j1,k1) = r_new(1,1);
        cyly_new(j1,k1) = r_new(2,1);
        cylz_new(j1,k1) = r_new(3,1);
    end
end

% Draw first half of bond with color1
surf(cylx_new+x(j),cyly_new+y(j), cylz_new...
+z(j),'facecolor',color1,'edgecolor',...
'none');
% Draw second half of bond with color2
surf(cylx_new+x(j)+segment_vector(1,1),...
cyly_new+y(j)+segment_vector(2,1),...
cylz_new+z(j)+segment_vector(3,1),...
'facecolor',color2,'edgecolor','none');
    end
  end
 end
end
end
```

The use of this function to draw a water molecule is given by these commands:

```
hold on
```

FIGURE 10.8 A H₂O molecule rendered with the draw_spheres and draw_bonds_cylinder functions.

```
draw_spheres('H2O-geo','O',[1 0 0],.05,40)
draw_spheres('H2O-geo','H',[.8 .8 .8],.03,40)
draw_bonds_cylinder('H2O-geo','O',[1 0 0],'H',...
   [.8 .8 .8],0.1,1.0,.03);
light
axis equal
axis off
view([1 -.5 0]);camup([0 0 -1]);
```

which can be combined into a script with a descriptive name, such as draw_H2O.m. The output of these commands is shown in Figure 10.8.

CHAPTER 11

MOs of the Trigonal Planar Geometry

INTRODUCTION

In the previous chapter, we saw how group theory can help us make sense of how the electronic structure of a water molecule emerges from the interactions of its atomic orbitals. Let's now explore how these ideas apply to a case with a more complicated arrangement: The trigonal planar geometry (D_{3h} symmetry). This case will make the utility of the projection operator method much more evident, as it provides a guide to handling multidimensional irreducible representations. As a specific example, we'll focus on a flattened ammonia molecule, which will later serve as our starting point for discussing how the shapes of molecules are influenced by electron counts.

CONSTRUCTION OF NH₃ GEOMETRIES

As always, our first step is to construct the geometry of the molecule, expressing the atomic positions in terms of Cartesian coordinates. The three-fold symmetry of the trigonal planar geometry means that obtaining these coordinates will require a bit of trigonometry. In anticipation of our future work on comparing the trigonal planar geometry to two alternatives (trigonal pyramidal and T-shaped), it is helpful to write a function for calculating the coordinates in terms of a few geometrical parameters: The N-H bond distance (d_NH), the angle between one N-H bond and the other two bonds when viewed down the three-fold axis (theta), and the angle by which the N-H bonds deviate from a planar geometry (phi).

A function for calculating these coordinates, outputting them to a file, and plotting the geometry is given here:

```
function NH3_geo(d_NH, phi, theta)

N_xyz = [0 0 0];
H1_xyz = [0 d_NH*cos(phi) d_NH*sin(phi)];
H2_xyz = [d_NH*sin(theta)*cos(phi) ...
d_NH*cos(phi)*cos(theta) d_NH*sin(phi)];
H3_xyz = [-d_NH*sin(theta)*cos(phi) ...
d_NH*cos(phi)*cos(theta) d_NH*sin(phi)];

f2=fopen('XH3-geo','w');
fprintf(f2,'N %f %f %f\n',N_xyz(1,1),N_xyz(1,2),...
N_xyz(1,3));
fprintf(f2,'H %f %f %f\n',H1_xyz(1,1),...
H1_xyz(1,2),H1_xyz(1,3));
fprintf(f2,'H %f %f %f\n',H2_xyz(1,1),...
H2_xyz(1,2),H2_xyz(1,3));
fprintf(f2,'H %f %f %f\n',H3_xyz(1,1),...
H3_xyz(1,2),H3_xyz(1,3));
fclose(f2);

figure

draw_bonds_cylinder('XH3-geo','N',[100 100 100]/...
255,'H',[250 250 250]/255,0.9*d_NH,1.1*d_NH,0.08);

axis off
axis equal
light
```

Exercise 11.1. Examine the above function and work through the following problems: (a) What units should d_NH, theta, and phi be provided in for the function to generate meaningful geometries that will be readable by our previous functions, such as build_hamiltonian? (b) What are the correct values of theta and phi for the trigonal planar, trigonal pyramidal, and T-shaped geometries? Make pictures of these geometries with the above function to check your answers (and test that the program is entered correctly).

MOs AT SPECIFIC GEOMETRIES

Once we have the above program ready, it is then simple to calculate the MOs and energies as a function of the geometrical parameters. We first generate the geometry for specific values of d_NH, theta, and phi. The geometry file is then loaded into build_hamiltonian to obtain the Hamiltonian matrix. Finally, the Hamiltonian matrix is diagonalized and the MOs are visualized. This process can be streamlined as follows:

```
function MO_E=solve_XH3(d_NH, phi, theta,mode)

NH3_geo(d_NH,phi,theta)

[H,S,orblist]=build_hamiltonian('XH3-geo');

[psi,E]=eig(H);

if(mode == 1)
for j=1:7
 figure
 set(gcf,'color',[1 1 1]);
 draw_bonds_cylinder('XH3-geo','N',[100 100 ...
     100]/255,'H',[250 250 250]/255,0.9*...
     d_NH,1.1*d_NH,0.08);
 axis off
 axis equal
 light
 drawMO('XH3-geo', orblist, psi(:,j), 2);
 view([.1,.1,1]);camup([0,1,0]);camzoom(1.8)
end
end

for j=1:sum(orblist)
MO_E(j,1) = E(j,j)
end
```

In this function, we've introduced a new feature: A variable (mode) that acts as a switch to turn on or off the execution of a portion of the code. The mode variable is given as an input parameter, and is used by the if(mode == 1) line to decide whether or not to carry out the lines between it and its corresponding end command. If the user provides 1 as the value of this variable, the seven MOs for the molecule are drawn. Otherwise, this part is skipped and just the MO energies are given as output.

Exercise 11.2. **The MO diagram of planar NH₃ calculated by brute force.** Plot the MO diagram for an NH_3 molecule with a trigonal planar geometry with a N-H bond distance of 1.00 Å (d_NH=1.00, theta=120*pi/180, and phi=0.0). You should obtain the energies and MOs shown in Figure 11.1.

In the MO diagram for a planar version of NH_3 obtained in Exercise 11.2, both familiar and unusual features can be noted. The bottom three MOs have clear N-H bonding overlap using the N 2s and in-plane N 2p orbitals, while the highest three are the N-H antibonding counterparts. These sets are consistent with the expectation that the molecule should have three N-H bonds. In addition, there is the unhybridized, nonbonding p orbital pointing perpendicular to the plane of the molecule. Altogether, these aspects appear to coincide with the notion that the molecule is sp²-hybridized.

The forms of some of the MOs, however, may seem less understandable, particularly if this is your first time seeing them. For example, why do the two MOs at −13.00 eV have the same energy, forming a degenerate pair,

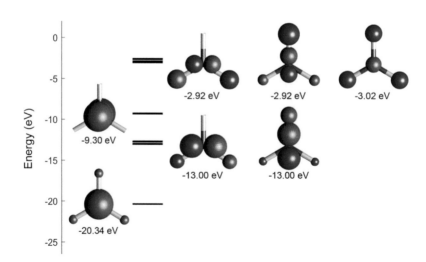

FIGURE 11.1 The MO diagram of a (hypothetical) planar NH_3 molecule.

when their shapes look so different? In the next sections, it will become clear how this degeneracy emerges through symmetry. At the same time, we will demonstrate the enormous utility of the projection operator method in symmetrizing a basis set.

SALCs FOR THE TRIGONAL PLANAR GEOMETRY

With a little help from group theory, the MO diagram of planar NH_3 becomes easy to derive. We begin by considering the basis set of atomic orbitals for the molecule, and how they transform under the symmetry operations of the molecule's D_{3h} point group. This basis set consists of the N 2s and 2p orbitals, as well as one 1s orbital for each of the H atoms (Figure 11.2). The N 2s orbital is symmetric with respect to all of the molecule's symmetry operations, so that it transforms as the totally symmetric representation of the point group. Upon referring to the D_{3h} character table (Table 11.1), we see that this representation is labelled as A_1'. The N $2p_z$ orbital can be similarly classified. It is symmetric with respect to the $\hat{E}$, $\hat{C}_3$, and $\hat{\sigma}_v$ operations, while being antisymmetric with respect to the $\hat{S}_3$, $\hat{C}_2'$, and $\hat{\sigma}_h$ operations (which interconvert +z and –z). It thus transforms as the A_2'' representation.

The classification of the remaining orbitals is not as obvious. Notice that rotating the N $2p_x$ does not correspond to simple multiplication by ±1, but instead results in a new p orbital rotated by 120° relative to the starting point. As p orbitals rotate the same way as vectors, this new orbital can be

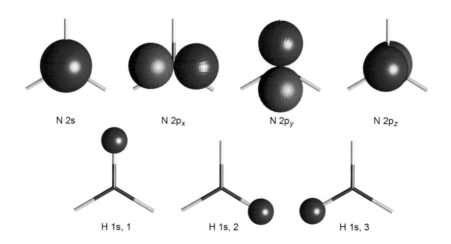

FIGURE 11.2   The basis set of atomic orbitals for a planar NH_3 molecule.

simply expressed as $\cos(120°)p_x + \sin(120°)p_y$, meaning that the p_x and p_y orbitals are mixed with each other by the rotation. The p_y orbital likewise becomes $-\sin(120°)p_x + \cos(120°)p_y$ upon rotation, such that the rotation can be represented with a matrix of the form:

$$\begin{pmatrix} c'_{px} \\ c'_{py} \end{pmatrix} = \begin{pmatrix} \cos(120°) & \sin(120°) \\ -\sin(120°) & \cos(120°) \end{pmatrix} \begin{pmatrix} c_{px} \\ c_{py} \end{pmatrix}$$

Neither the p_x or p_y orbitals on their own can represent D_{3h} symmetry, but instead must be considered as two members of a 2D representation. From the character table, we can see that there are two such irreducible representations for this point group: E′ and E″. By noting that the E′ representation is symmetric with respect to the reflection through the molecular plane, while E″ is antisymmetric, we can assign the p_x and p_y orbitals to the E′ representation.

Exercise 11.3. **Characters for the p_x/p_y representation.** Write 2×2 matrices describing how the p_x and p_y orbitals transform under each of the symmetry operations of the D_{3h} point group. Confirm that the traces of the matrices correspond to the characters of the E′ irreducible representation in Table 11.1.

We now come to the H 1s orbitals. As with the p_x and p_y orbitals, the H 1s orbitals individually do not form representations of the point group symmetry. Instead many of the symmetry operations interconvert the H 1s orbitals such that we should write 3×3 matrices to represent the operations. As the D_{3h} character table does not have any 3D irreducible representations, this representation must be reducible, and we should decompose it into irreducible parts before attempting to build an MO diagram. In Chapter 9,

TABLE 11.1   Character Table the Point Group D_{3h}

D_{3h}	$\hat{E}$	$2\hat{C}_3$	$2\hat{C}_2'$	$\hat{\sigma}_h$	$2\hat{S}_3$	$3\hat{\sigma}_v$
A_1'	1	1	1	1	1	1
A_2'	1	1	-1	1	1	-1
E′	2	-1	0	2	-1	0
A_1''	1	1	1	-1	-1	-1
A_2''	1	1	-1	-1	-1	1
E″	2	-1	0	-2	1	0

we built up just the functions we need for carrying out this reduction: `reduce_representation` and `projection_operator`.

To start we first enter in the necessary information for the D_{3h} point group. The character table can be typed in as follows:

```
>> Characters_D3h = [1  1  1   1   1   1 % A_1'
                     1  1 -1   1   1  -1 % A_2'
                     2 -1  0   2  -1   0 % E'
                     1  1  1  -1  -1  -1 % A_1''
                     1  1 -1  -1  -1   1 % A_2''
                     2 -1  0  -2   1   0 % E''
                                          ];
```

In addition, we need a separate array to specify the number of operations of each class in the point group:

```
>> D3_class_mult = [1 2 3 1 2 3];
```

where these numbers are simply read from the column headers in Table 11.1. This defines the "character vector space" in which the reducible representation is going to be decomposed into its irreducible components. As a final step to this definition, we give names to the irreducible representations:

```
>> D3h_names = ['a1''  '; 'a2''  '; 'e''';  ...
      'a1'''''; 'a2''''''; 'e''''  ']
```

where we have used the fact that MATLAB® considers two single quotes in a row as a single quote in the string (whereas a single quote on its own denotes the start or end of string). Also note that care was taken to ensure that all of the names have the same number of characters.

Next we need to determine the characters for the matrices representing the symmetry operations in the basis of the three H 1s orbitals. For the $\hat{E}$ and $\hat{\sigma}_h$ operations, all three H 1s orbitals are left in place, making the characters 3. The $\hat{C}_3$ and $\hat{S}_3$ operations, on the other hand, all permute the H 1s orbitals, leaving none in the same place. The characters for these operations are then 0. Finally, the $\hat{\sigma}_v$ and $\hat{C}_2'$ operations interconvert two of the H 1s orbitals, while leaving one unchanged, resulting in a character of 1. These conclusions are entered in as:

```
>> Characters_Hs = [3 0 1 3 0 1];
```

With this preliminary work done, we can then use reduce_
representation to project the character vector for the H 1s representation onto the basis vectors for the character vector space defined by the
irreducible representations:

```
>> reduce_representation(Characters_Hs,...
   Characters_D3h,D3_class_mult,D3h_names)
Group order = 12
Reducible representation =+ 1*a1'+ 1*e'
```

From this, we have determined that basis of three H 1s orbitals can be
reduced to one A_1' function (matching the symmetry of the N 2s orbital),
and a pair of functions transforming as an E' representation (matching
the N $2p_x$ and $2p_y$ orbitals).

To finish the symmetry analysis, we now need to generate these reduced
functions. We have already built the code to carry out this process in
the projection_operator function. Before we can use it, though,
we need to explicitly enter in the 12 matrices describing how the three
H 1s orbitals are transformed under each of the symmetry operations.
Fortunately, all of the matrix elements are just 1's and 0's:

```
>> D3h_Hs(:,:,1)  =  [1    0    0;
                      0    1    0;
                      0    0    1];
>> D3h_Hs(:,:,2)  =  [0    0    1   %C_3
                      1    0    0
                      0    1    0];
>> D3h_Hs(:,:,3)  =  D3h_Hs(:,:,2)*D3h_Hs(:,:,2);
       %C_3^2
>> D3h_Hs(:,:,4)  =  [1    0    0   %C_2 #1
                      0    0    1
                      0    1    0];
>> D3h_Hs(:,:,5)  =  [0    0    1   %C_2 #2
                      0    1    0
                      1    0    0];
>> D3h_Hs(:,:,6)  =  [0    1    0   %C_2 #3
                      1    0    0
                      0    0    1];
>> D3h_Hs(:,:,7)  =  [1    0    0   %sigma_h
                      0    1    0
                      0    0    1];
```

```
>> D3h_Hs(:,:,8) = [0    0    1%S3
                    1    0    0
                    0    1    0];
>> D3h_Hs(:,:,9) = D3h_Hs(:,:,8)*D3h_Hs(:,:,8);
    %S_3^2
>> D3h_Hs(:,:,10) = [1   0    0  %sigma_v#1
                     0   0    1
                     0   1    0];
>> D3h_Hs(:,:,11) = [0   0    1  %sigma_v#2
                     0   1    0
                     1   0    0];
>> D3h_Hs(:,:,12) = [0   1    0  %sigma_v#3
                     1   0    0
                     0   0    1];
```

Once these matrices are set up, we can then use the projection operator approach to start with the first H 1s orbital, and build up linear combinations that transform as the designated irreducible representations. For the A_1' function, the command is:

```
>> SALC_a1 = projection_operator([1,0,0]', ...
   D3h_Hs, [1 1 1 1 1 1 1 1 1 1 1 1]);
```

where the $[1,0,0]'$ vector gives the coefficients for each of the three 1s orbitals in the starting point, and the row of 12 ones lists the A_1' characters for each of the 12 symmetry operations in the point group (not just for one representative of each class, as is done in the standard character table).

To visualize the result, we should first set up the corresponding vectors in the full atomic orbital basis set of the molecule:

```
>> Hs_ao = [0 0 0
0 0 0
0 0 0
0 0 0
1 0 0
0 1 0
0 0 1];
```

After that, the SALC_a1 vector can then be converted to the full basis set using the basis2ao function with basis2ao(SALC_e1_1, Hs_ao), and plotted with drawMO. Since we will be plotting many functions in a similar way, it is convenient to make quick function to draw the orbitals along with the molecular geometry for any given set of orbital coefficients:

```
function drawMO_NH3(psi)
  orblist= [4 1 1 1];
  figure
  set(gcf,'color',[1 1 1]);
  draw_bonds_cylinder('XH3-geo','N',[100 100 100]/...
    255,'H',[250 250 250]/255,0.9,1.1,0.04);
  axis off
  axis equal
  light
  drawMO('XH3-geo', orblist, psi, 2);
  view([.1,.1,1]);camup([0,1,0]);camzoom(2.8);
```

The result of following these steps for the A'_1 function is shown in Figure 11.3. It consists of a superposition of the H 1s orbitals with equal phases and weights. A quick inspection of function confirms that it is symmetric with respect to all symmetry operations of the molecule, and thus indeed transforms as the A'_1 irreducible representation. Notice also that we will obtain the same function regardless of what starting point we used: [1,0,0]', [0,1,0]' or [0,0,1]' (as you can verify).

Such is not the case for the E' functions. When we enter in the commands:

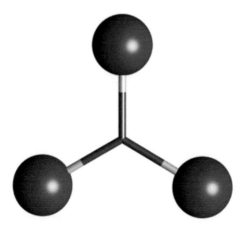

FIGURE 11.3   The A'_1 combination of H 1s orbitals.

```
>> SALC_e1_1 = projection_operator([1,0,0]', ...
   D3h_Hs, [2 -1 -1 0 0 0 2 -1 -1 0 0 0]);
>> SALC_e1_2 = projection_operator([0,1,0]', ...
   D3h_Hs, [2 -1 -1 0 0 0 2 -1 -1 0 0 0]);
>> SALC_e1_3 = projection_operator([0,0,1]', ...
   D3h_Hs, [2 -1 -1 0 0 0 2 -1 -1 0 0 0]);
>> drawMO_NH3(basis2ao(SALC_e1_1,Hs_ao));camzoom(2)
>> drawMO_NH3(basis2ao(SALC_e1_2,Hs_ao));camzoom(2)
>> drawMO_NH3(basis2ao(SALC_e1_3,Hs_ao));camzoom(2)
```

the three functions displayed in Figure 11.4 are the result. In each case, the H 1s contribution at the starting point is twice as big and of opposite sign from the remaining two H 1s contributions. At first glance, it might appear strange that we have three different functions for a representation that is supposed to be only two dimensional. This issue is resolved, however, by noting that these functions are not all linearly independent of each other. Try plotting the sum of the first two functions, and you will find that the result is proportional to the third function. There are only two unique functions here from which the third is a linear combination.

Even if we just take, say, the first two functions, however, they do not make a convenient representation in their current form. In particular, they are not orthogonal to each other, as can be seen by calculating their overlap:

```
>> SALC_e1_1'*SALC_e1_2

ans =

 -0.5000
```

A much more useful set can be obtained by orthogonalizing the functions to each other. **Schmidt Orthogonalization** is one approach. In this

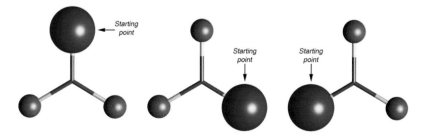

FIGURE 11.4   Functions generated for the E′ representation using three different starting points.

method, an orthogonal set is made from two non-orthogonal functions, ψ_1 and ψ_2, by supplementing ψ_2 with the right amount of ψ_1 to make it orthogonal to ψ_1, i.e. $|\psi_2'\rangle = |\psi_2\rangle + c|\psi_1\rangle$, with $\langle \psi_2' | \psi_1 \rangle = 0$. Applying $\langle \psi_1 |$ to both sides of the former equation gives $\langle \psi_1 | \psi_2' \rangle = \langle \psi_1 | \psi_2 \rangle + c \langle \psi_1 | \psi_1 \rangle$, which simplifies to $c = -\langle \psi_1 | \psi_2 \rangle$.

In MATLAB, this process is carried out as follows:

```
>> SALC_e1_2_orth = SALC_e1_2 - (SALC_e1_1'*...
   SALC_e1_2)*SALC_e1_1;
>> SALC_e1_2_orth = SALC_e1_2_orth/...
   (SALC_e1_2_orth'*SALC_e1_2_orth)^0.5;
```

And the orthogonality of the functions can be confirmed by typing SALC_e1_1'*SALC_e1_2_orth.

The final E' functions are presented in Figure 11.5, alongside the A_1' SALC derived earlier. Now the symmetry match with the N p_x and p_y is much clearer. The first E' function has a node running horizontally along the page through the molecule, which aligns closely with the nodal plane of the N $2p_y$ orbital. The second E' function has a node running vertically, aligning with that of the N $2p_x$ orbital.

BUILDING THE MO DIAGRAM FROM THE SALCs

Now that we have finished symmetrizing the basis set of atomic orbitals, the construction of the MO diagram is straightforward. Within the basis set, we have two A_1' functions, two sets of E' functions, and one A_2' function. The interactions between the two A_1' functions give rise to N-H bonding and antibonding MOs at −20.34 eV and −3.02 eV, respectively. The two sets of E' functions also yield bonding and antibonding combinations, this

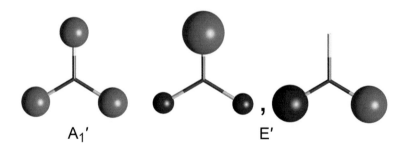

FIGURE 11.5 Symmetry adapted linear combinations (SALCs) of the H 1s orbitals in planar NH_3.

time at −13.00 eV and −2.92 eV, respectively. Finally, the A_2' N $2p_z$ orbital has no functions of the same symmetry to interact with, and so remains nonbonding at −9.30 eV.

Exercise 11.4. Create interaction diagrams analogous to that shown in Figure 10.4 for the A_1' and E′ interactions described above.

CHAPTER 12

Walsh Diagrams and Molecular Shapes

INTRODUCTION

Up until now, we have focused on the construction of MO diagrams for molecules in fixed geometries. One of the key applications of electronic structure theory, however, is in making predictions of the preferred geometry of a system of atoms, and providing explanations for these preferences in terms of chemical bonding. In this chapter, we will illustrate one approach to making these predictions: The use of **Walsh diagrams**. A Walsh diagram is a plot that charts the MO energies of a molecule as we distort it from one geometry to another. From following how the energy levels change under the distortion, we can evaluate whether the perturbation of the geometry should be favorable or unfavorable. In general, the outcome will depend on the electron count of the molecule, allowing us to develop rules relating geometry to electron count.

GEOMETRIES OF THE AL₃ MOLECULE

To illustrate how Walsh diagrams can be constructed and applied, we'll use the NH_3 molecule from the last chapter as a model system. We can consider it to represent the more general situation of a main group atom (A) bound to three ligand groups (L) which interact with atom A in a σ fashion. For such AL_3 molecules, three geometries are common: Trigonal pyramidal (as in NH_3, PH_3, etc.), trigonal planar (BF_3), and T-shaped (ClF_3).

To determine the relative stabilities of these geometries using Walsh diagrams, the first step is to draw pathways between them. Group theory provides one way to organize these paths. The D_{3h} point symmetry of the trigonal planar geometry is a **supergroup** of those of the trigonal pyramidal (C_{3v}) and the T-shaped (C_{2v}) geometries, meaning that it contains all of the symmetry operations of these lower symmetry **subgroups**. In other words, the trigonal pyramidal and T-shaped geometries can be obtained by breaking symmetry operations of the trigonal planar geometry. The C_{3v} and C_{2v} point groups, however, are not connected by such a supergroup-subgroup relationship. In that sense, it is simplest to start with the trigonal planar geometry, whose MO diagram we just derived, and watch how its MO energies change as we distort it.

The necessary distortions can be easily accomplished by changing the θ and ϕ angles that we encoded previously in the `solve_NH3` function away from the values ($\phi = 0°$, $\theta = 120°$). Changing ϕ while holding θ at 120° corresponds to the creation of the trigonal pyramidal geometry, where the σ_h mirror plane and C_2-axes perpendicular to the C_3-axis are lost. On the other hand, shifting θ from 120° to 90° (while holding ϕ at 0°) opens one of the L-A-L bond angles to create the T-shaped geometry. Here only one C_2-axis, the σ_h mirror plane, and the σ_v plane running parallel to the maintained C_2-axis survive the distortion.

CONSTRUCTING WALSH DIAGRAMS

To create a Walsh diagram for one of these distortions, all we need to do is carry out MO energy calculations for a series of steps along the distortion path. For example, on going from trigonal planar to the trigonal pyramidal geometry, we might consider changing ϕ from 0° to 30° in steps of 1°. This series can be encoded as a `for` loop of the form "`for phi = 0:30` ... `end`", with each iteration of the loop containing a calculation of the MOs for the current θ angle.

One point to consider here is how to save the MO energies that are obtained in a way that will be easy to plot later. As the `solve_NH3` function returns the MO energies for the geometry it is given as a column vector, the most convenient way to arrange these energies for a series of geometries is to line up these column vectors into a matrix, with one column for each θ angle. The nth row of the matrix then corresponds to how the energy of the nth MO changes during the distortion.

A script implementing this procedure can be constructed as follows:

```
phi_count=0;
for phi = 0:30
  phi_count=phi_count+1;
  E(:,phi_count)=solve_XH3(1.0,phi*pi/180,120...
    *pi/180,2);
end
```

Once this has finished running, the rows of the matrix MO energies can be simply plotted one-by-one as functions of angle:

```
figure
hold on
for j=1:7
  plot(0:30,E(j,:),'color',[0,0,0]);
end
```

Exercise 12.1. Use the above script to recreate the Walsh Diagram shown in Figure 12.1 for the distortion of an AL_3 molecule from the trigonal planar geometry to the trigonal pyramidal one. To obtain pictures of the MOs at

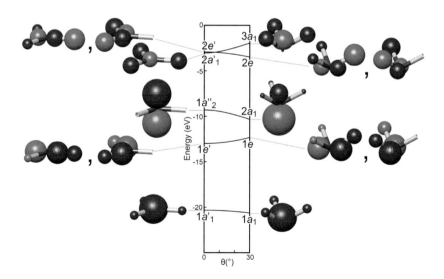

FIGURE 12.1   Walsh diagram for the distortion of an AL_3 molecule from trigonal planar (left) to trigonal pyramidal (right).

the end points, the `solve _ NH3` function can be run separately for these geometries with the `mode` value set to 1.

On the left-hand side of the diagram in Figure 12.1, the MOs we obtained in the last chapter for the trigonal planar geometry appear: The lowest A s-based bonding MO ($1a_1'$ = the first MO that transforms according to the a_1' irreducible representation of D_{3h} symmetry), followed by a degenerate pair of bonding MOs centered by the A p_x and p_y orbitals ($1e'$), the nonbonding A p_z orbital ($1a_2''$), and finally the antibonding levels ($2a_1'$ and $2e''$). As we move from left to right across the diagram, these energies begin to change, with some MO energies going down and others going up. As the various MOs disagree on whether this geometrical change is favorable, the geometry ultimately adopted by the molecule will depend on the occupation of these levels. The most straightforward approach to investigating any particular electron configuration would be to sum the energies of the electrons to obtain a total energy, and see which angle minimizes it.

A more qualitative guideline for judging the favorability of the distortion is given by **Walsh's rule**: The geometry adopted by a molecule will be that which best stabilizes its highest occupied molecular orbital (HOMO) or nearby occupied MOs. Through this principle, we can see that at two electrons and eight electrons the pyramidal geometry would be preferred, in keeping with general preference for condensed clusters at low electron counts and the observed shape of octet molecules such as NH_3, respectively. A six-electron molecule, however, is expected to stay planar, as is the case for PF_3 and would be the case for BH_3 were it not to dimerize to make B_2H_6 molecules.

At ten electrons, the HOMO for the trigonal planar molecule goes up upon pyramidalization, crossing the $2e'$ level as it comes down from slightly higher energies (an artifact of the antibonding levels being bunched together at moderately high energy in Hückel results, underestimating the $2a_1' - 2e'$ energy gap). Moving to a pyramidal geometry does not seem to be downhill at ten electrons, and populating the diagram up to the next gap at 14 electrons will effectively remove any net bonding in the molecule.

Exercise 12.2. Now that we've explored the pyramidal geometry, let's move to the third arrangement: T-shaped. Modify the above script to create a Walsh diagram for the distortion of a trigonal planar molecule to a T-shaped one. You should be able to obtain the diagram shown in Figure 12.2.

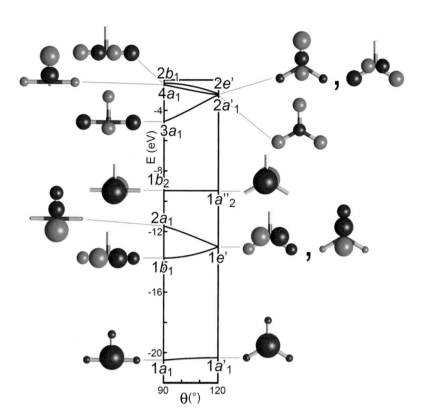

FIGURE 12.2   Walsh diagram following the MO energies for an AL_3 molecule as it goes between a trigonal planar geometry (right) and a T-shaped one (left).

Walsh's rule also allows the rationalization of the electron counts that prefer the T-shaped geometry. Six-electron and eight-electron AL_3 molecules are expected to resist distortions toward the T-shaped geometry, consistent with their experimental preference for trigonal planar and trigonal pyramidal geometries, respectively. Ten-electron molecules, however, would show a strong driving force for adopting the T-shaped geometry as the original $2a_1'$ antibonding MO becomes the more nonbonding $3a_1$ MO.

CHAPTER 13

Getting Started with Transition Metals

INTRODUCTION

To discuss inorganic compounds containing transition metals, it will be important to be able to include d orbitals in our analyses. Putting in this functionality will, of course, require us to adapt some of our MATLAB® functions to accommodate this larger basis set. In addition, we will also need to incorporate the more complicated basis functions that are used to describe d orbitals. They are generally described not by a single Slater-type orbital, but by a linear combination of two Slater-type functions with different ζ values:

$$\psi_{n,l,m} = c_1 \left(\frac{(2\zeta_1)^{n+1/2}}{((2n)!)^{1/2}} \right) r^{n-1} e^{-\zeta_1 r} \cdot Y_{l,m}(\theta,\phi) + c_2 \left(\frac{(2\zeta_2)^{n+1/2}}{((2n)!)^{1/2}} \right) r^{n-1} e^{-\zeta_2 r} \cdot Y_{l,m}(\theta,\phi)$$

where one of the ζ values is usually relatively large (to model to the short-range behavior of the orbital) and the other is small (to capture the long-range behavior). c_1 and c_2 are coefficients that allow us to adjust the relative contributions of the short-range and long-range components. These types of functions are known as *double-zeta* functions. In this chapter, we will develop this capability in our MATLAB code, and apply it to calculating the MO diagram of a hypothetical octahedral CrH_6^{6-} anion, a model of an 18-electron transition metal complex with σ-ligands.

NORMALIZATION OF DOUBLE-ZETA FUNCTIONS

The values of c_1 and c_2 are generally given as parameters for a Hückel or extended Hückel calculation. They must be chosen with care, however, so that the overall double-zeta function is normalized. The parameters for the Cr atom we will use for CrH_6^{6-}, for example, are listed in a set of DFT-calibrated Hückel parameters as:

$H_{ii}(4s) = -4.721$ eV $\quad \zeta_{4s} = 2.3368$

$H_{ii}(4p) = -2.350$ eV $\quad \zeta_{4p} = 2.1417$

$H_{ii}(3d) = -8.829$ eV $\quad \zeta_{3d,1} = 6.0897 \quad \zeta_{3d,2} = 2.3300$
$\quad\quad\quad\quad\quad\quad\quad\quad\quad\quad\quad c_1 = 1.000 \quad\quad c_2 = 4.3318$

which we can specify as a MATLAB array as:

```
[4 -4.721   2.3368  4  -2.350   2.1417 3 -8.829...
   6.0897   2.3300  1.000  4.3318];
```

From their magnitudes, we can see that the c_1 and c_2 values here are just giving the correct c_1/c_2 ratio rather than the magnitudes needed to create a normalized d orbital.

To obtain the correction magnitudes for c_1 and c_2, we can use MATLAB to calculate the overlap of the d orbital with itself:

```
function [c1, c2] = double_zeta_norm(atom_pos1,...
coeff1,STO_parameters)

atom_pos1 = atom_pos1/0.5291772;  % convert
                                  % Angstroms to Bohr.
%    n_d   = STO_parameters(7);
%    Hii_d = STO_parameters(8);
%    zeta_d1 = STO_parameters(9);
%    zeta_d2 = STO_parameters(10);
%    c1 = STO_parameters(11);
%    c2 = STO_parameters(12);

STO_parameters1 = STO_parameters;
STO_parameters1(12) = 0;
STO_parameters2 = STO_parameters;
STO_parameters2(11) = 0;
```

```
% <psi|psi> = <c1*zeta1|c1*zeta1> + 2<c1*zeta1|
            % c2*zeta2> + <c2*zeta2|c2*zeta2>
psi1 = @(x,y,z) STO_psi(x,y,z,atom_pos1,...
       STO_parameters1, coeff1);
psi2 = @(x,y,z) STO_psi(x,y,z,atom_pos1,...
       STO_parameters2, coeff1);

Sintegrand = @(x,y,z) psi1(x,y,z).*psi2(x,y,z);
zeta1_zeta2 = integral3(Sintegrand,-Inf,Inf,...
              -Inf,Inf,-Inf,Inf,'AbsTol',0.000001);

Sintegrand = @(x,y,z) psi1(x,y,z).*psi1(x,y,z);
zeta1_zeta1 = integral3(Sintegrand,-Inf,Inf,...
              -Inf,Inf,-Inf,Inf,'AbsTol',0.000001);

Sintegrand = @(x,y,z) psi2(x,y,z).*psi2(x,y,z);
zeta2_zeta2 = integral3(Sintegrand,-Inf,Inf,...
              -Inf,Inf,-Inf,Inf,'AbsTol',0.000001);

% <psi|psi> = <zeta1|zeta1> + 2<zeta1|zeta2> + ...
            % <zeta2|zeta2>

psi_psi = zeta1_zeta1 + 2*zeta1_zeta2 + zeta2_zeta2;

c1 = STO_parameters(11)/sqrt(psi_psi);
c2 = STO_parameters(12)/sqrt(psi_psi);
```

Exercise 13.1. Annotate the above MATLAB function to explain how it obtains normalized values for c_1 and c_2. Use the function to calculate such values for the d orbitals of Cr.

INCLUSION OF D ORBITALS INTO MATLAB FUNCTIONS

The next step in the incorporation of d orbitals into our analysis is to adapt our MATLAB functions to work with them. Here are updated versions of build_hamiltonian, STO_psi, and STO_Hij_Sij:

```
function [H_ao, S_ao, orb_list] = ...
   build_hamiltonian(filename)

[atomname,x,y,z] = textread(filename,'%s %f %f %f');
natoms = size(x);
```

```matlab
natoms = natoms(1);

atom_pos = [x y z];

params = zeros(natoms,6+2+4);
orb_list = ones(1,natoms);
% Assign STO parameters
num_orbitals = 0;
for j = 1:natoms
    foundit = 0;
    if(strcmp(atomname(j),'H')==1)
            params(j,1:6) = [1 -6.692 2.5981 0 0.0 ...
                              0.0];
            num_orbitals = num_orbitals+1;
            orb_list(1,j) = 1;
            foundit = 1;
    end
    if(strcmp(atomname(j),'C')==1)
            params(j,1:6) = [2 -11.2580 ...
                    2.25  2  -7.8540  2.2173];
            num_orbitals = num_orbitals+4;
            orb_list(1,j) = 4;
            foundit = 1;
    end
    if(strcmp(atomname(j),'N')==1)
            params(j,1:6) = [2 -16.507 2.455 ...
                      2  -9.300  2.397];
            num_orbitals = num_orbitals+4;
            orb_list(1,j) = 4;
            foundit = 1;
    end
    if(strcmp(atomname(j),'O')==1)
            params(j,1:6) = [2 -23.487 2.5055 ...
                      2  -9.177  1.8204];
            num_orbitals = num_orbitals+4;
            orb_list(1,j) = 4;
            foundit = 1;
    end
    if(strcmp(atomname(j),'Cr')==1)
        params(j,:) = ...
        [4 -4.721   2.3368  4  -2.350   2.1417 3 ...
            -8.829   6.0897  2.3300 0.20521  0.88902];
```

```
            num_orbitals = num_orbitals+9;
            orb_list(1,j) = 9;
            foundit = 1;
        end

    if(foundit == 0)
        fprintf('Parameters for atom %d not
          found.\n',j);
    end
end

num_orbitals

% CALCULATION H AND S MATRICES
H_ao = zeros(num_orbitals, num_orbitals);
S_ao = zeros(num_orbitals, num_orbitals);
atomj = 1;
for j = 1:num_orbitals
    fprintf('Interactions for orbital
            %d\n',j);
    % Determine whether we've already gone through
    % all orbitals on atom;
    if j > sum(orb_list(1,1:atomj))
        atomj = atomj + 1;
    end
    % Set position of nucleus orbital sits on.
    atom_pos1 = atom_pos(atomj,:);

    coeff1 = zeros(1,9);
    % Determine which orbital on atom we are
    % dealing with.
    ao_num = j;
    if (atomj > 1)
        ao_num = j - sum(orb_list(1,1:(atomj-1)));
    end
    coeff1(ao_num) = 1;

    atomk = 1;
    for k = 1:num_orbitals
        if k > sum(orb_list(1,1:atomk))
            atomk = atomk + 1;
        end
```

```
        % Set position of nucleus orbital sits on.
        atom_pos2 = atom_pos(atomk,:);
        coeff2 = zeros(1,9);
        % Determine which orbital on atom we are
        % dealing with.
        ao_num = k;
        if (atomk > 1)
            ao_num = k - sum(orb_list(1,1:(atomk-1)));
        end
        coeff2(ao_num) = 1;
        if(j==k)
          if(ao_num == 1)
             H_ao(j,j) = params(atomj,2);
             S_ao(j,j) = 1;
          end
          if(ao_num>1)&&(ao_num<5)
             H_ao(j,j) = params(atomj,5);
             S_ao(j,j) = 1;
          end
          if(ao_num>4)&&(ao_num<10)
             H_ao(j,j) = params(atomj,8);
             S_ao(j,j) = 1;
          end
        else
          if(atomj~=atomk)  % Don't waste time
            % calculating overlaps on same atom.
            [H_ao(j,k) S_ao(j,k)] =
              STO_Hij_Sij(atom_pos1,coeff1,...
        params(atomj,:),atom_pos2,coeff2,...
        params(atomk,:));
          end
        end
    end
end

function psi = STO_psi(x,y,z,atom_pos, ...
  STO_parameters, coefficients)
%
%    atom_pos = [Atomic number, x_nucleus, y_
%    nucleus, z_nucleus]
%    STO_parameters = [n_s Hii_s  zeta_s   n_p
%    Hii_p  zeta_p ]
%    coefficients  = [c_s c_px c_py c_pz]
```

```
%

x_rel = x - atom_pos(1);
y_rel = y - atom_pos(2);
z_rel = z - atom_pos(3);

c_s  = coefficients(1);
c_px = coefficients(2);
c_py = coefficients(3);
c_pz = coefficients(4);
n_s    = STO_parameters(1);
Hii_s = STO_parameters(2);
zeta_s = STO_parameters(3);
n_p    = STO_parameters(4);
Hii_p = STO_parameters(5);
zeta_p = STO_parameters(6);


[phi, theta_matlab, r] = cart2sph(x_rel,y_rel,z_rel);
theta = pi/2 - theta_matlab;

 % Define normalization constant, N
 N_s = (2*zeta_s)^(n_s+0.5)/(factorial(2*n_s))^0.5;
 N_p = (2*zeta_p)^(n_p+0.5)/(factorial(2*n_p))^0.5;
 % Calculate value of radial wavefunction
   R_s = N_s*r.^(n_s-1).*exp(-zeta_s*r);
   R_p = N_p*r.^(n_p-1).*exp(-zeta_p*r);

Y_s = 1/(2*pi^0.5);
Y_px = (3/(4*pi))^0.5.*sin(theta).*cos(phi);
Y_py = (3/(4*pi))^0.5.*sin(theta).*sin(phi);
Y_pz = (3/(4*pi))^0.5.*cos(theta);


psi = c_s*R_s.*Y_s + c_px*R_p.*Y_px + c_py*R_p.*...
   Y_py + c_pz*R_p.*Y_pz;

ncoefficients = size(coefficients);
if(ncoefficients(2) > 4)
   c_dx2y2 = coefficients(5);
   c_dz2   = coefficients(6);
   c_dxy   = coefficients(7);
```

```
    c_dxz  = coefficients(8);
    c_dyz = coefficients(9);
    n_d = STO_parameters(7);
    Hii_d = STO_parameters(8);
    zeta_d1 = STO_parameters(9);
    zeta_d2 = STO_parameters(10);
    c1 = STO_parameters(11);
    c2 = STO_parameters(12);
    N_d1 = (2*zeta_d1)^(n_d+0.5)/...
            (factorial(2*n_d)) ^0.5;
    N_d2 = (2*zeta_d2)^(n_d+0.5)/...
            (factorial(2*n_d)) ^0.5;
    R_d = c1*N_d1*r.^(n_d-1).*exp(-zeta_d1*r)+c2*...
        N_d2*r.^(n_d-1).*exp(-zeta_d2*r);
    Y_dx2y2=(15/(16*pi))^0.5*sin(theta).*sin(theta)....
            *cos(2*phi);
    Y_dz2=0.25*(5/(pi))^0.5*(3*cos(theta).*...
          cos(theta)-ones(size(theta)));
    Y_dxz=0.5*(15/(pi))^0.5*(sin(theta).*...
          cos(theta).*cos(phi));
    Y_dyz=0.5*(15/(pi))^0.5*(sin(theta).*...
          cos(theta).*sin(phi));
    Y_dxy=(15/(16*pi))^0.5*sin(theta).*...
          sin(theta).*sin(2*phi);
    psi = psi + ...
      c_dx2y2*R_d.*Y_dx2y2+c_dz2*R_d.*Y_dz2+c_dxy*...
      R_d.*Y_dxy+c_dxz*R_d.*Y_dxz+c_dyz*R_d.*Y_dyz;
end

function [H_ij, S_ij] = ...
STO_Hij_Sij(atom_pos1,coeff1,STO_params1,...
  atom_pos2,coeff2, STO_params2)

atom_pos1 = atom_pos1/0.5291772;  % convert
                % Angstroms to Bohr.
atom_pos2 = atom_pos2/0.5291772;  % convert
                % Angstroms to Bohr.

psi1 = @(x,y,z) STO_psi(x,y,z,atom_pos1,...
       STO_params1, coeff1);
```

```
psi2 = @(x,y,z) STO_psi(x,y,z,atom_pos2,...
        STO_params2, coeff2);
Sintegrand = @(x,y,z) psi1(x,y,z).*psi2(x,y,z);
S_ij = integral3(Sintegrand,-Inf,Inf,-Inf,Inf,...
        -Inf,Inf,'AbsTol',0.0001);

K = 1.75; % Wolfsberg-Helmholz constant.

% STO_parameters = [n_s H_ii(s) zeta_s n_p...
                    % H_ii(p) zeta_p]
H_ii_s = STO_params1(1,2);
H_ii_p = STO_params1(1,5);
H_ii_1 = norm(coeff1(1))^2*H_ii_s + ...
         (norm(coeff1(2))^2+norm(coeff1(3))^2+...
         norm(coeff1(4))^2)*H_ii_p;

H_ii_d = STO_params1(1,8);
H_ii_1 = H_ii_1 + (norm(coeff1(5))^2+...
         norm(coeff1(6))^2+ ...
         norm(coeff1(7))^2+norm(coeff1(8))^2+...
         norm(coeff1(9))^2)*H_ii_d;

H_ii_s = STO_params2(1,2);
H_ii_p = STO_params2(1,5);
H_ii_2 = norm(coeff2(1))^2*H_ii_s + ...
         (norm(coeff2(2))^2+norm(coeff2(3))^2+...
         norm(coeff2(4))^2)*H_ii_p;

H_ii_d = STO_params1(1,8);
H_ii_2 = H_ii_2 + (norm(coeff2(5))^2+...
         norm(coeff2(6))^2+ ...
         norm(coeff2(7))^2+norm(coeff2(8))^2+...
         norm(coeff2(9))^2)*H_ii_d;

H_ij = K*0.5*(H_ii_1+H_ii_2)*S_ij;
```

Exercise 13.2. Examine the revised functions above, and explain the role of each of the changes from the originals marked in black.

THE MOs OF AN OCTAHEDRAL COMPLEX WITH σ-LIGANDS; THE 18-ELECTRON RULE

Let's now try out this code on an octahedral CrH_6^{6-} anion, as a model for transition metals with σ-ligands. We'll use the following atomic coordinates:

```
Cr    0.000000    0.000000    0.000000
 H    1.110000    0.000000    0.000000
 H   -1.110000    0.000000    0.000000
 H    0.000000    1.110000    0.000000
 H    0.000000   -1.110000    0.000000
 H    0.000000    0.000000    1.110000
 H    0.000000    0.000000   -1.110000
```

Before running the calculation, it is important to judge the realism of our parameters for Cr and H. As the hydrogen in this case is expected to be rather anionic, the H parameters we used before (optimized for H_2O) are not ideal. In particular, note from the code above that the H_{ii} for the H 1s is listed as being above the Cr 3d. To qualitatively correct for this, let's lower the H 1s orbital and make it less contracted, using the following parameters: $H_{ii}(1s) = -9.50$ eV, $\zeta(1s) = 2.00$ (playing with these parameters can give you valuable information about how the bonding the molecule depends to the character of the H atom orbitals).

Exercise 13.3. **Debugging your code and obtaining the basic MO diagram.** Run a Hückel calculation on this molecule with these updated parameters. If all is well, you should obtain the following MO energies:

```
>> Energies=eig(H_CrH6);
for j=1:15
fprintf('MO %d:E = %f eV\n',j,Energies(j));
end
MO 1:   E = -20.629101 eV
MO 2:   E = -17.558831 eV
MO 3:   E = -17.558696 eV
MO 4:   E = -14.051277 eV
MO 5:   E = -14.051242 eV
MO 6:   E = -14.051218 eV
MO 7:   E = -8.829000 eV
MO 8:   E = -8.829000 eV
```

```
MO 9:    E = -8.829000 eV
MO 10:   E = 0.750797 eV
MO 11:   E = 0.750922 eV
MO 12:   E = 2.325297 eV
MO 13:   E = 2.325319 eV
MO 14:   E = 2.325356 eV
MO 15:   E = 2.993674 eV
```

Note that a very large energy gap occurs between MO's 9 and 10, suggesting that a favorable closed-shell electron configuration would be achieved at a filling of 18 electrons. The origin of this gap can be clearly seen by plotting pictures of the MOs, as is done in Figure 13.1 on the next page. In the lowest six orbitals, the strong bonding overlap appears between the central Cr atom and the neighboring H atoms, while the remaining three are simply nonbonding Cr d orbitals. Upon moving to the tenth MO, however, Cr-H antibonding occurs: The gap between MO 9 and 10 then coincides with the filling of all bonding and nonbonding MOs, and the onset of antibonding.

It is no accident that this gap occurs at this particular electron count. Note that each of the nine MOs below the gap is centered by a different Cr valence atomic orbital. In this way, filling all of these levels would correspond to having an electron pair associated with each of the Cr atom's valence orbitals, in much the same way that placing eight electrons around a carbon atom allows to make full use of its four s and p orbitals for bonding. This expectation of special stability at 18 electrons for a transition metal atom is known as the **18-electron rule**, which can be evoked to rationalize the inertness or thermodynamic stability of molecules that adhere to it, or the reactivity of those that violate it.

Exercise 13.4. Plot the MOs of CrH_6 as is done in Figure 13.1.

If we look closely at the MOs with degenerate energies, we may notice that their orbital contributions do not seem to align very well with the symmetry elements of the molecule. For example, MOs 8 and 9 might be expected to consist of d orbitals lying in mirror planes perpendicular to the C_4-axes of the molecule to form the d orbital set that transforms as the t_{2g} irreducible representation of the system's O_h point group symmetry (d_{xy}, d_{xz}, and d_{yz}). Similarly, the Cr orbitals centering MOs 2 and 3 look as though they are approximating the d_{z2} and d_{x2-y2} orbitals, but they lack the C_4 symmetry along the z-axis that would be expected for these functions.

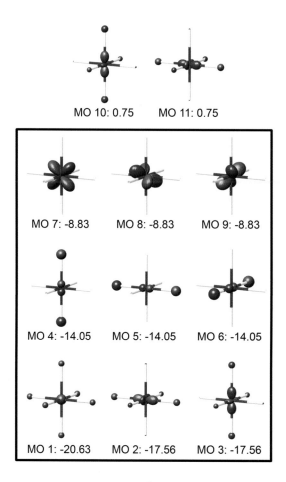

MO 10: 0.75   MO 11: 0.75

MO 7: -8.83   MO 8: -8.83   MO 9: -8.83

MO 4: -14.05   MO 5: -14.05   MO 6: -14.05

MO 1: -20.63   MO 2: -17.56   MO 3: -17.56

FIGURE 13.1 The lowest 11 MOs of CrH_6^{6-}, as obtained from the diagonalization of the Hückel Hamiltonian. Note that the first nine levels are Cr-H bonding or Cr nonbonding, and are each centered by a different Cr s, p, or d valence atomic orbital. This association of an electron pair with each of the valence orbitals of a transition metal atom is referred to as an 18-electron configuration.

These unusual shapes enter in due to the freedom offered by the energy degeneracies of these levels to MATLAB's matrix diagonalization algorithm. When several MOs are eigenvectors of the Hamiltonian with the same energy, any linear combination of them will also yield eigenvectors with the same energy. The solutions that MATLAB produces are not necessarily the set that aligns with the axes of our coordinate system. To obtain clear representations of degenerate sets of MOs, we may then need to take linear combinations of the MATLAB solutions.

To illustrate this process, let's consider the set defined by MOs 2 and 3. The Cr centers in these two functions bear orbitals that are mixtures of the Cr d_{z2} and d_{x2-y2}. To see how these two contributors can be disentangled, we print out the atomic orbital coefficients for the MOs:

```
>> for j=1:9
fprintf('Atomic Orbital %d:%f (MO 2)%f(MO 3) \n',...
          j, psi_CrH6(j,2), psi_CrH6(j,3));
end
Atomic Orbital 1:   0.000002 (MO 2) -0.000005 (MO 3)
Atomic Orbital 2:   0.000000 (MO 2) -0.000000 (MO 3)
Atomic Orbital 3:   0.000000 (MO 2) -0.000000 (MO 3)
Atomic Orbital 4: -0.000000 (MO 2)  0.000000 (MO 3)
Atomic Orbital 5:   0.695526 (MO 2) -0.198641 (MO 3)
Atomic Orbital 6:   0.198640 (MO 2)  0.695526 (MO 3)
Atomic Orbital 7:   0.000000 (MO 2)  0.000000 (MO 3)
Atomic Orbital 8:   0.000000 (MO 2) -0.000000 (MO 3)
Atomic Orbital 9:   0.000000 (MO 2) -0.000000 (MO 3)
>>
```

where psi_ CrH6 is the matrix whose columns are the atomic orbital coefficients of the MOs. Your numbers may vary from these due to the above-mentioned degree of freedom in the matrix diagonalization. If so, simply adapt the procedure below to your specific numerical values.

From this chart, we see that Cr contribution to MO 2 consists mostly of the d_{x2-y2} orbital, but that d_{z2} also appears with a coefficient of 0.198640. MO 3, meanwhile, has a large component from d_{z2}, but that some d_{x2-y2} sneaks in with a coefficient of -0.198641, a value essentially equal in magnitude to the d_{z2} impurity in MO 2. With this information, we are ready to produce a new set of functions of the form: $\psi_1' = \psi_1 + \delta\psi_2$ and $\psi_2' = \psi_2 - \delta\psi_1$, where the equal value of δ for the two functions ensures that they stay orthogonal.

This process can be carried out as follows:

```
>> psi_CrH6_orth = psi_CrH6;
>> psi_CrH6_orth(:,2) = psi_CrH6(:,2) - ...
      (0.198640/0.695526)*psi_CrH6(:,3);
>> psi_CrH6_orth(:,3) = psi_CrH6(:,3) + ...
      (0.198640/0.695526)*psi_CrH6(:,2);
>> psi_CrH6_orth(:,2) = psi_CrH6_orth(:,2)/...
      norm(psi_CrH6_orth(:,2));
```

```
>> psi_CrH6_orth(:,3) = psi_CrH6_orth(:,3)/...
    norm(psi_CrH6_orth(:,3));
```

Executing this code then yields the functions shown in Figure 13.2. The new set consists of one MO centered by the Cr d_{z2} orbital and another centered by the d_{x2-y2} orbital.

The three-fold degeneracy of MOs 7–9 makes finding a perfect resolution of three d_{xy}, d_{xz}, and d_{yz} components less straightforward using such pair-wise linear combinations. While linear algebra could be applied to get an exact solution, the correct qualitative picture can be obtained by simply recombining MOs 8 and 9 as shown in Figure 13.3. Of course, since these functions are non-interacting with the H atoms, and simply consist of linear combinations of the d_{xy}, d_{xz}, and d_{yz} orbitals, we could manually construct the well-oriented functions as:

```
>> psi_CrH6_orth(:,7:9) = 0;
>> psi_CrH6_orth(7,7) = 1;
>> psi_CrH6_orth(8,8) = 1;
>> psi_CrH6_orth(9,9) = 1;
```

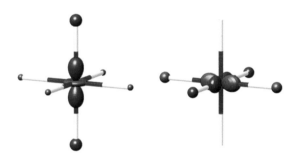

FIGURE 13.2 New combinations of MOs 2 and 3 to separate the d_{z2} and d_{x2-y2} components of the set.

FIGURE 13.3 New combinations of MOs 7–9 to better separate their d_{xy}, d_{xz}, and d_{yz} components.

CHAPTER 14

Translational Symmetry and Band Structures

INTRODUCTION

Earlier, we saw how symmetry places restrictions on the wavefunctions of a system: A molecule's energy eigenfunctions must transform as irreducible representations of the point group symmetry, and interactions between functions transforming as different irreducible representations are forbidden by symmetry. These properties greatly simplified the construction of MO diagrams for molecules. The power of group theory, however, is even more evident in the way that it opens a path to calculate the electronic structure of periodic structures, such as crystals and surfaces—systems that contain a nearly limitless number of atoms, and would thus lead to unmanageably large Hamiltonian matrices without the help of symmetry. In this chapter, we will see how the application of symmetry to such structures leads to the concepts of k-space and energy bands, allowing us to expand our MATLAB® code to work with crystalline materials.

TRANSLATIONAL SYMMETRY AND BLOCH'S THEOREM

Periodic structures have symmetry operations that we have not considered so far: lattice translations, in which a shift of the structure by certain vectors leads to a structure indistinguishable from the original. For a 3D crystal, these vectors are simply all of the possible linear combinations of the $\mathbf{a}$, $\mathbf{b}$, and $\mathbf{c}$ vectors defining the unit cell with integer coefficients, as follows:

$$\mathbf{R}_{n_a n_b n_c} = n_a \mathbf{a} + n_b \mathbf{b} + n_c \mathbf{c}$$

From these vectors, we can define a series of operations corresponding to the full set of lattice translations:

$$\hat{T}_{n_a n_b n_c} = \hat{T}_{\mathbf{a}}^{n_a} + \hat{T}_{\mathbf{b}}^{n_b} + \hat{T}_{\mathbf{c}}^{n_c}$$

Just as for rotations, reflections, and inversions, the electron density should be left invariant to each of these operations, as is expressed in the following equation:

$$\hat{T}_{n_a n_b n_c} \rho(\mathbf{r}) = \rho(\mathbf{r} - n_a \mathbf{a} - n_b \mathbf{b} - n_c \mathbf{c}) = \rho(\mathbf{r})$$

where $\rho(\mathbf{r})$ is the electron density of the system. This invariance of the electron density with respect to translations also restricts the form of the wavefunctions. For example, if we were to consider a single electron in the periodic system, we would have:

$$\hat{T}_{n_a n_b n_c} \psi(\mathbf{r}) = \psi'(\mathbf{r}), \text{ with } |\psi(\mathbf{r})|^2 = |\psi'(\mathbf{r})|^2$$

such that the square of the original and translated wavefunctions are equal, giving the same electron density. Some simple answers to the form of $\psi'(\mathbf{r})$ are $\psi'(\mathbf{r}) = +\psi(\mathbf{r})$ or $\psi'(\mathbf{r}) = -\psi(\mathbf{r})$, i.e. the wavefunction could be symmetric or antisymmetric with respect to translations. For operations of order 2, such as mirror operations where $\hat{\sigma}^2 = \hat{E}$, these are the only two possibilities. Lattice translations, however, can be repeated any number of times, creating many more valid relationships between $\psi(\mathbf{r})$ and $\psi'(\mathbf{r})$. In general, any of equations of the form:

$$\hat{T}_{n_a n_b n_c} \psi(\mathbf{r}) = e^{i\theta} \psi(\mathbf{r})$$

would satisfy the symmetry restrictions, as:

$$|\psi'(\mathbf{r})|^2 = \left| e^{i\theta} \psi(\mathbf{r}) \right|^2 = e^{-i\theta} \psi^*(\mathbf{r}) \cdot e^{i\theta} \psi(\mathbf{r}) = \psi^*(\mathbf{r})\psi(\mathbf{r}) = |\psi(\mathbf{r})|^2$$

In this way, the wave functions of a periodic structure transform one-dimensional representations of the lattice translations with characters of the form $e^{i\theta}$.

To go further with our discussion of translational symmetry, it is useful to set up somewhat artificial boundary conditions on the system, known as **periodic boundary conditions**. Here, we put bounds on the number of distinct translations in the system by stating that after a given number of shifts along **a**, **b**, and **c** we return to the original point, that is:

$$\hat{T}_\mathbf{a}^{N_a} = \hat{T}_\mathbf{b}^{N_b} = \hat{T}_\mathbf{c}^{N_c} = \hat{E}$$

where N_a, N_b, and N_c are integers. This statement can be made relatable with a geographic analogy: The Earth appears flat to an observer on the ground, and treating it as such is perfectly reasonable for making measurements on a short length-scale. However, if that observer were to travel far enough, that very tired individual would return to the point of departure. Imagining the topology of the situation for periodicity along three dimensions is a little tricky, but the size of N_a, N_b, and N_c can be made so large that the boundaries become an insignificant part of the system. In fact, we will eventually consider the limit where each of these numbers go to infinity.

Once, the periodic boundary conditions are in place, the wavefunctions can be understood in greater detail. As $\hat{T}_\mathbf{a}^{N_a} = \hat{T}_\mathbf{b}^{N_b} = \hat{T}_\mathbf{c}^{N_c} = \hat{E}$, we have a series of equations restricting the characters for each of the lattice translations. For example, for translations along $\mathbf{a}$, we can write:

$$\hat{T}_\mathbf{a}^{N_a}\psi(\mathbf{r}) = e^{iN_a\theta_a}\psi(\mathbf{r}) = e^{i2\pi m_a}\psi(\mathbf{r}) = \psi(\mathbf{r})$$

in which θ_a is the change in phase-angle that the wavefunction undergoes upon being translated once along $\mathbf{a}$ and m_a is an integer. From this equation, the possible values of θ_a are given by:

$$\theta_a = \frac{2\pi m_a}{N_a} = 2\pi q_a$$

with q_a being a semi-continuous variable, giving rise to N_a distinct values for $e^{i\theta_a}$ between $m_a = 0$ (where $e^{i\theta_a} = 1$) and $m_a = N_a$ (where $e^{i\theta_a}$ is again 1). When we make similar considerations for the $\mathbf{b}$ and $\mathbf{c}$ axes of the crystal, we find that the wavefunction should transform in the following way with respect to any given set of translations:

$$\hat{T}_{n_a n_b n_c}\psi(\mathbf{r}) = e^{i2\pi(n_a q_a + n_c q + n_c q_c)}\psi(\mathbf{r})$$

The phase shifts between neighboring lattice points in the crystal have a very close relationship to the differences in phase of incident light being scattered from these points during a diffraction experiment. For this reason, it is common to discuss the phase-angles in terms of the **reciprocal lattice** used for interpreting diffraction patterns. The reciprocal lattice is defined as being the lattice of points created from linear combinations of three vectors, **a***, **b***, and **c*** related to **a**, **b**, and **c**, through the equations:

$$\mathbf{a}^* = \frac{2\pi\, \mathbf{b}\times\mathbf{c}}{\mathbf{a}\cdot(\mathbf{b}\times\mathbf{c})}, \quad \mathbf{b}^* = \frac{2\pi\, \mathbf{c}\times\mathbf{a}}{\mathbf{a}\cdot(\mathbf{b}\times\mathbf{c})}, \quad \mathbf{c}^* = \frac{2\pi\, \mathbf{a}\times\mathbf{b}}{\mathbf{a}\cdot(\mathbf{b}\times\mathbf{c})}$$

such that **a*** is orthogonal to **b** and **c**, the dot product of **a*** and **a** is 2π (meaning that **a*** becomes shorter as **a** becomes longer), and **b*** and **c*** exhibit analogous properties. Using these definitions, we can then rewrite $2\pi q_a$ as $k_x\,\mathbf{a}^*\cdot\mathbf{a}$, $2\pi q_b$ as $k_y\,\mathbf{b}^*\cdot\mathbf{b}$, and $2\pi q_c$ as $k_z\,\mathbf{c}^*\cdot\mathbf{c}$. The phase shifts upon translating the wavefunction then take the form:

$$\hat{T}_{n_a n_b n_c}\,\psi(\mathbf{r}) = e^{i\mathbf{k}\cdot(n_a\mathbf{a}+n_b\mathbf{b}+n_c\mathbf{c})}\psi(\mathbf{r})$$

where **k** is a vector in an abstract space referred to as **momentum space** (since functions of the form $e^{i\mathbf{k}\cdot\mathbf{r}}$ are eigenvectors for the quantum mechanical momentum operator with eigenvalues of $\hbar\mathbf{k}$), **reciprocal space**, or simply k-space. This conclusion is codified in **Bloch's theorem** for periodic structures, which states that a one-electron wavefunction in such a structure can be written as $\psi(\mathbf{r}) = e^{i\mathbf{k}\cdot\mathbf{r}}\,\chi(\mathbf{r})$, where $\chi(\mathbf{r})$ is a periodic function with $\chi(\mathbf{r}+n_a\mathbf{a}+n_b\mathbf{b}+n_c\mathbf{c}) = \chi(\mathbf{r})$.

The above conclusions gain immense power when we connect them to the concepts of group theory that we discussed in previous chapters. In particular, if we consider $\psi(\mathbf{r})$ as a representation of the lattice symmetry of the structure, the phase factors $e^{i\mathbf{k}\cdot(n_a\mathbf{a}+n_b\mathbf{b}+n_c\mathbf{c})}$ can be recognized as the characters for the $\hat{T}_{n_a n_b n_c}$ operations in this representation. In this way, values of **k** yielding different values of $e^{i\mathbf{k}\cdot(n_a\mathbf{a}+n_b\mathbf{b}+n_c\mathbf{c})}$ correspond to different irreducible representations of the lattice symmetry, which are forbidden by symmetry to interact with each other.

CONSTRUCTING SALCs

The identification of **k** with irreducible representations of the translational symmetry allows us to symmetrize the basis set of a crystalline material. Recall that symmetry adapted linear combinations $\psi_{\text{SALC, irred }i}$ of basis functions ϕ_j can be created using the projection operator method:

$$\psi_{\text{SALC, irred }i} = \left(\sum_{\hat{R}} \chi^*_{\text{irred }i}(\hat{R})\,\hat{R}\right)\phi_{\text{starting point}}$$

The same construction can be used for translational symmetry. Let's consider a crystal whose basis set consists of the atomic orbital set $\{\phi_{j,n_a n_b n_c}\}$, where j runs over the atomic orbitals within a single unit cell, while $n_a n_b n_c$ indicates which unit cell the orbital is in. Symmetrized functions transforming as irreducible representations of the lattice translations are constructed as:

$$\phi_j(\mathbf{k}) = \frac{1}{\sqrt{N}} \sum_{n_a} \sum_{n_b} \sum_{n_c} e^{-i\mathbf{k}\cdot(n_a\mathbf{a}+n_b\mathbf{b}+n_c\mathbf{c})} \phi_{j,n_a n_b n_c}$$

where N is the number of unit cells within the edges of the periodic boundary conditions, i.e. $N = N_a \times N_b \times N_c$. We then build such symmetrized functions for each atomic orbital in the unit. As only functions with the same $\mathbf{k}$ value can interact with each other, we have reduced the problem of considering an infinite number of atomic orbitals interacting with each other to a limited number of functions interacting (one for each orbital in the unit cell) at many different $\mathbf{k}$ points.

HAMILTONIAN MATRICES

We are almost ready to start putting together MATLAB code for calculating the electronic structure of crystalline materials. What remains is to develop equations for the Hamiltonian matrix elements representing the interactions between the $\phi_j(\mathbf{k})$ functions. These matrix elements can be evaluated as follows:

$$H_{jk}(\mathbf{k}) = \langle \phi_j(\mathbf{k}) | \hat{H} | \phi_k(\mathbf{k}) \rangle$$

$$= \frac{1}{N} \left(\sum_{n_a,n_b,n_c} e^{i\mathbf{k}\cdot(n_a\mathbf{a}+n_b\mathbf{b}+n_c\mathbf{c})} \langle \phi_{j,n_a n_b n_c} | \hat{H} \left(\sum_{m_a,m_b,m_c} e^{-i\mathbf{k}\cdot(m_a\mathbf{a}+m_b\mathbf{b}+m_c\mathbf{c})} | \phi_{k,m_a m_b m_c} \rangle \right) \right)$$

$$= \frac{1}{N} \sum_{n_a,n_b,n_c} \sum_{m_a,m_b,m_c} e^{i\mathbf{k}\cdot((n_a-m_a)\mathbf{a}+(n_b-m_b)\mathbf{b}+(n_c-m_c)\mathbf{c})} \langle \phi_{j,n_a n_b n_c} | \hat{H} | \phi_{k,m_a m_b m_c} \rangle$$

Notice that the terms in the sum only depend on the relative placement of unit cells defined by the indices $n_a n_b n_c$ and $m_a m_b m_c$. We can thus rewrite the $m_a m_b m_c$ indices in terms of their differences from their counterparts in the $n_a n_b n_c$ indices: $m_a \to m_a - n_a = \Delta n_a$, $m_b \to m_b - n_b = \Delta n_b$, and $m_c \to m_c - n_c = \Delta n_c$. The Hamiltonian matrix elements then become:

$$H_{jk}(\mathbf{k}) = \frac{1}{N} \sum_{n_a} \sum_{n_b} \sum_{n_c} \sum_{\Delta n_a} \sum_{\Delta n_b} \sum_{\Delta n_c} e^{-i\mathbf{k}\cdot(\Delta n_a\mathbf{a}+\Delta n_b\mathbf{b}+\Delta n_c\mathbf{c})} \langle \phi_{j,000} | \hat{H} | \phi_{k,\Delta n_a \Delta n_b \Delta n_c} \rangle$$

The sum over n_a, n_b, and n_c consists of N identical terms, such that the sum cancels the $1/N$ normalization factor, leading to a simple formula:

$$H_{jk}(\mathbf{k}) = \sum_{\Delta n_a}\sum_{\Delta n_b}\sum_{\Delta n_c} e^{-i\mathbf{k}\cdot(\Delta n_a\mathbf{a}+\Delta n_b\mathbf{b}+\Delta n_c\mathbf{c})} \left\langle \phi_{j,000}\left|\hat{H}\right|\phi_{k,\Delta n_a\Delta n_b\Delta n_c}\right\rangle$$

This equation has a straightforward interpretation. The Hamiltonian matrix element consists of the sum of interactions between atomic orbital j in the central unit cell and copies of atomic orbital k in the various unit cells, with each term having a phase factor that varies with $\mathbf{k}$.

The above expression for $H_{jk}(\mathbf{k})$ becomes even more manageable when we consider how the interaction between atomic orbitals will vary with Δn_a, Δn_b, and Δn_c. As the magnitudes of these numbers increase, the k atomic orbital is placed further and further away from the j atomic orbital in the central cell, leading to a smaller interaction. The series of $\Delta n_a = 0, \pm 1, \pm 2, \pm 3, \ldots$ can then be safely truncated at a $\Delta n_{a,max}$ value that corresponds to negligible interactions. For example, values such that $\Delta n_{a,max}{\cdot}a$, $\Delta n_{b,max}{\cdot}b$, and $\Delta n_{c,max}{\cdot}c > 10$ Å would safely capture all chemically significant interactions, and setting the thresholds to >5 Å would probably lead to a virtually indistinguishable result. The best approach, of course, is to make sure that the results of your calculation are converged with respect to these limits.

A SIMPLE EXAMPLE: THE CHAIN OF H ATOMS

After these many pages of derivations, we are in great need of a tangible example. Let's start with a classic model system for periodic structures, a 1D chain of hydrogen atoms arranged with a repeat vector $\mathbf{a}$ (see Hoffmann, R. *Solids and Surfaces: A Chemist's View of Bonding in Extended Structures*; VCH Publishers: New York, NY, 1988.). In this case, each unit cell contains only a single hydrogen 1s orbital, the Hamiltonian matrix consists of a single element that is a function of $\mathbf{k}$:

$$H_{11}(\mathbf{k}) = \sum_{\Delta n_a} e^{-i\mathbf{k}\cdot\Delta n_a\mathbf{a}} \left\langle \phi_{H1s,0}\left|\hat{H}\right|\phi_{H1s,\Delta n_a}\right\rangle$$

We can then expand the sum and group the terms according to how far away each $\phi_{H1s,\Delta n_a}$ orbital is from the central unit cell:

$$H_{11}(\mathbf{k}) = \left\langle \phi_{H\,1s,0} \middle| \hat{H} \middle| \phi_{H\,1s,0} \right\rangle + \left(e^{-i\mathbf{k}\cdot\mathbf{a}} + e^{i\mathbf{k}\cdot\mathbf{a}} \right)\left\langle \phi_{H\,1s,0} \middle| \hat{H} \middle| \phi_{H\,1s,\pm1} \right\rangle$$

$$+ \left(e^{-i\mathbf{k}\cdot2\mathbf{a}} + e^{i\mathbf{k}\cdot2\mathbf{a}} \right)\left\langle \phi_{H\,1s,0} \middle| \hat{H} \middle| \phi_{H\,1s,\pm2} \right\rangle + \cdots$$

where we have used the fact since the s-s orbital interaction depends only on distance $\left\langle \phi_{H\,1s,0} \middle| \hat{H} \middle| \phi_{H\,1s,+n} \right\rangle = \left\langle \phi_{H\,1s,0} \middle| \hat{H} \middle| \phi_{H\,1s,-n} \right\rangle = \left\langle \phi_{H\,1s,0} \middle| \hat{H} \middle| \phi_{H\,1s,\pm n} \right\rangle$. This expression can be simplified further by recalling the Euler relation $e^{i\theta} + e^{-i\theta} = 2cos\theta$ to:

$$H_{11}(\mathbf{k}) = H_{ii,1s} + 2\cos(\mathbf{k}\cdot\mathbf{a})\left\langle \phi_{H\,1s,0} \middle| \check{H} \middle| \phi_{H\,1s,\pm1} \right\rangle$$

$$+ 2\cos(\mathbf{k}\cdot2\mathbf{a})\left\langle \phi_{H\,1s,0} \middle| \check{H} \middle| \phi_{H\,1s,\pm2} \right\rangle + \cdots$$

or, if we simply limit ourselves to interactions between neighboring unit cells:

$$H_{11}(\mathbf{k}) = H_{ii,1s} + 2\cos(\mathbf{k}\cdot\mathbf{a})\beta_{1s,1s} = H_{ii,1s} + 2\cos(k_x\cdot a)\beta_{1s,1s}$$

where $\beta_{1s,1s}$ is the interaction energy between two hydrogen 1s orbitals at a distance of a from each other.

Exercise 14.1. **The band structure of a hydrogen atom chain.** In Chapter 4, we calculated the simple Hückel Hamiltonian matrix for an H_2 molecule with a bond distance of 0.70 Å. Let's use the numerical values for $H_{ii,1s}$ (–7.528 eV) and $\beta_{1s,1s}$ (–7.976 eV) from that analysis to plot the energies of the wavefunctions for an H atom chain with $a = 0.70$ Å as a function of k. We begin by specifying the key parameters for the system:

```
H_ii = -7.528;
beta_1s1s = -7.976;
a = 0.70;
```

then calculate the energies for a range of k values. Note that for the cosine function maxima and minima will occur for k being multiples of π/a, so it makes sense to use this value unit in the range:

```
k = -3*pi/a:.01:3*pi/a;
E_k = H_ii + 2*beta_1s1s*cos(k*a);
```

Finally, we can plot the energies as a function of k. Below is the code for this, along with some extra lines to define the grid lines and mark the intervals of π/a along the x-axis of the plot:

```
plot(k,E_k,'color',[0,0,0],'linewidth',2);
hold on;
title('\fontsize{18}\cdot\cdot\cdotH\cdot\cdot...
    \cdotH\cdot\cdot\cdotH\cdot...
     \cdot\cdotH\cdot\cdot\cdotH\cdot\cdot\cdotH...
      \cdot\cdot\cdot');
ax = gca;
ax.XTick = [-3*pi/a -2*pi/a -pi/a 0 pi/a 2*pi/ ...
    a 3*pi/a];
ax.XTickLabel = {'-3\pi/a' '-2\pi/a' '-\pi/a' ...
    '\Gamma=0' '\pi/a' '2\pi/a' '3\pi/a'};
ax.XGrid = 'on';
ax.XLabel.String = 'k';
ax.YLabel.String = 'Energy (eV)';
plot([-3*pi/a 3*pi/a], [H_ii H_ii],':','color', ...
    [0,0,0],'linewidth',1);
```

Confirm that through using these commands you obtain the plot shown in Figure 14.1.

UNIQUE VALUES OF K: THE 1ST BRILLOUIN ZONE

In looking at the band structure we just obtained for the H atom chains (Figure 14.1), you probably noticed that the curve follows a repeating pattern, with points separated by intervals of $2\pi/a$, i.e. a reciprocal lattice vector, being equivalent. This follows simply from the form of the wavefunction:

$$\phi_j(k_x) = \frac{1}{\sqrt{N}} \sum_{n_a} e^{-i\left(k_x + \frac{2\pi}{a}\right)n_a a} \phi_{H1s,n_a} = \frac{1}{\sqrt{N}} \sum_{n_a} e^{-ik_x n_a a + 2\pi n_a} \phi_{H1s,n_a} = \frac{1}{\sqrt{N}} \sum_{n_a} e^{-ik_x n_a a} \phi_{H1s,n_a}$$

Shifting k_x by a reciprocal lattice vector spacing here simply leads to phase shifts of multiples of 2π, which leaves the overall wavefunction unchanged.

The same considerations apply to higher dimensional crystals as well, with the wavefunctions generated with $k+h\mathbf{a}^*+k\mathbf{b}^*+l\mathbf{c}^*$ being equivalent for any integer values of h, k, and l. All of the unique wavefunctions possible for a system will lie within one repeat unit of the reciprocal lattice.

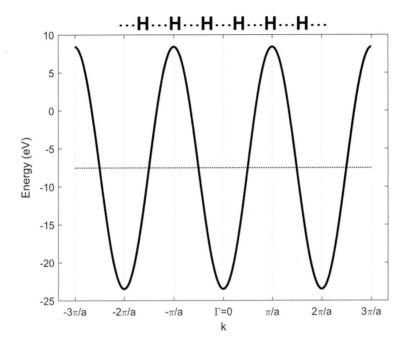

FIGURE 14.1   Simple Hückel band structure for a chain of H atoms with a spacing of $a = 0.7$ Å.

The traditional way to represent this domain of unique values of **k** is the take the volume of k-space closer to the **k** = 0 point, known as Γ, than to any other reciprocal lattice point. The resulting volume is known as the 1st Brillouin Zone (or often simply as the **Brillouin Zone**). To calculate the electronic structure of a compound, one carries out calculations of the wavefunctions for a grid of k-points designed to sample the Brillouin Zone.

BUILDING THE HAMILTONIAN MATRICES FOR PERIODIC STRUCTURES

Let's now adapt our Hückel code to generating Hamiltonian matrices for a more complicated crystal structure. The first issue is specifying the geometry of the structures. Whereas for a molecule we could simply supply a list of atoms and their coordinates, here we must provide the positions of the atoms in a periodic structure. To do this, we give separate files for the atoms within a unit cell and the repeat vectors for the unit cell. For example, the files for a layer of graphene might look like this:

```
File: graphene-geo
   C 0.00000 0.00000 0.00000
   C 0.00000 1.42610 0.00000
File: graphene-cell
  2.47000 0.00000 0.00000
 -1.23500 2.13908 0.00000
  0.00000 0.00000 106.79000
```

The file graphene-geo contains the positions of two C atoms separated by 1.43 Å in the y-direction, following our usual format. The file graphene-cell, meanwhile, contains the **a**, **b**, and **c** vectors as rows. The **a** and **b** vectors are of equal length and oriented 120° from each other to propagate a honeycomb pattern from the two C atoms in the –geo file. The **c** vector is set to an extremely long length to create an essentially 2D system; the third vector is included as the MATLAB code we develop below will expect a 3D periodic structure.

Now that we have established our how geometry files will be structured, we can begin the process of constructing the k-dependent Hamiltonian matrices. Note that in our above equations, the interactions between the irreducible representations for each atomic orbital will involve matrix elements between pairs of atomic orbitals in different unit cells. It will be useful to start by building up tables of the Hamiltonian and Overlap matrix elements between atomic orbitals with different Δn_a, Δn_b, and Δn_c values.

The function begins as usual with definitions of its input and output variables:

```
function [H_ao, S_ao, orb_list] = ...
  build_hamiltonian_trans(filename, overlaps)
```

where filename will be the stem for the –geo and –cell files, and the overlaps variable will be a vector with three integers corresponding to $\Delta n_{a,max}$, $\Delta n_{b,max}$, and $\Delta n_{c,max}$. H _ ao and S _ ao will be similar to the Hamiltonian and Overlap matrices we have calculated before, except for each pair of atomic orbitals separate entries will be made for each combination of Δn_a, Δn_b, and Δn_c values.

As the function runs, it will need to translate atoms relative to each other by the unit cell vectors. One of the first steps will then be to read in these vectors from the –cell file:

```
cellfile = strcat(filename,'-cell');
[cell_column1,cell_column2,cell_column3] ...
  = textread(cellfile,'%f %f %f');
cellmatrix = [cell_column1,cell_column2,...
  cell_column3];
```

where the rows of `cellmatrix` will be the **a**, **b**, and **c** vectors.

Now that we have loaded the cell vectors, we can calculate the Hamiltonian matrix elements for the interactions of the atoms in one unit cell with those displaced by different Δn_a, Δn_b, and Δn_c values. We do this by looping over these three variables over the range $\pm \Delta n_{a,max}$, $\pm \Delta n_{b,max}$, and $\pm \Delta n_{c,max}$, respectively:

```
overlap_counter=0;
for j1=-overlaps(1):overlaps(1)
  for j2=-overlaps(2):overlaps(2)
    for j3=-overlaps(3):overlaps(3)
      trans_vector=j1*cellmatrix(1,:)+j2*...
       cellmatrix(2,:)+j3*cellmatrix(3,:);
      [H_j1j2j3, S_j1j2j3, orb_list]= ...
      build_ hamiltonian_intercell ...
      (filename, trans_vector);
      overlap_counter=overlap_counter+1;
      H_ao(:,:,overlap_counter) = H_j1j2j3;
      S_ao(:,:,overlap_counter) = S_j1j2j3;
    end
  end
end
```

Here, `build_hamiltonian_intercell` is a yet-to-be written function that will introduce a shift of `trans_vector` between the atoms on the left- and right-hand sides of the H_{ij} matrix elements in our original `build_hamiltonian` function. Also note that we have numbered the various Δn_a, Δn_b, and Δn_c combinations with the counter `overlap_counter`. As long as we always use the same loop order to scan over these indices, no ambiguity will arise here.

The full function should then be of the form:

```
function [H_ao, S_ao, orb_list] = ...
  build_hamiltonian_trans(filename, overlaps)
cellfile = strcat(filename,'-cell');
```

```
[cell_column1,cell_column2,cell_column3] = ...
  textread(cellfile,'%f %f %f');
cellmatrix = [cell_column1,cell_column2,...
  cell_column3];
overlap_counter=0;
for j1=-overlaps(1):overlaps(1)
  for j2=-overlaps(2):overlaps(2)
    for j3=-overlaps(3):overlaps(3)
      trans_vector=j1*cellmatrix(1,:)+j2*...
      cellmatrix(2,:)+j3*cellmatrix(3,:);
       [H_j1j2j3, S_j1j2j3, orb_list]=...
       build_ hamiltonian_intercell...
       (filename, trans_vector);
       overlap_counter=overlap_counter+1;
       H_ao(:,:,overlap_counter) = H_j1j2j3;
       S_ao(:,:,overlap_counter) = S_j1j2j3;
    end
  end
end
```

Exercise 14.2. **The build_hamiltonian_intercell function.** Use the build_hamiltonian function from the previous chapters as the basis for the creation of the build_hamiltonian_intercell function to fulfill the needs indicated in the program above.

Exercise 14.2 Solution. The modified portions of build_hamiltonian are indicated in black in the code below.

```
function [H_ao, S_ao, orb_list] = ...
  build_ hamiltonian_intercell(filename, trans_vector)

geoname = strcat(filename,'-geo');
[atomname,x,y,z] = textread(geoname,'%s %f %f %f');

natoms = size(x);
natoms = natoms(1);

atom_pos = [x y z];

params = zeros(natoms,6+2+4);
orb_list = ones(1,natoms);
% Assign STO parameters
```

```
num_orbitals = 0;
for j = 1:natoms
  foundit = 0;
  % H from Benzene
  if(strcmp(atomname(j),'H')==1)
      params(j,1:6) = [1 -9.500 2.00 0 0.0 0.0];
      num_orbitals = num_orbitals+1;
      orb_list(1,j) = 1;
      foundit = 1;
  end
  if(strcmp(atomname(j),'C')==1)
      params(j,1:6) = [2 -11.2580 2.25 2 -7.8540 ...
        2.2173];
      num_orbitals = num_orbitals+4;
      orb_list(1,j) = 4;
      foundit = 1;
  end
  if(strcmp(atomname(j),'N')==1)
      params(j,1:6) = [2 -16.507 2.455 2 ...
        -9.300 2.397];
      num_orbitals = num_orbitals+4;
      orb_list(1,j) = 4;
      foundit = 1;
  end
  if(strcmp(atomname(j),'O')==1)
    params(j,1:6) = [2 -23.487 2.5055 2 ...
      -9.177 1.8204];
    num_orbitals = num_orbitals+4;
    orb_list(1,j) = 4;
    foundit = 1;
  end
  if(strcmp(atomname(j),'Cr')==1)
    params(j,:) = [4 -4.721 2.3368 4 -2.350 ...
      2.1417 3 -8.829 6.0897 2.3300 0.20521 0.88902];
    num_orbitals = num_orbitals+9;
    orb_list(1,j) = 9;
    foundit = 1;
  end

  if(foundit == 0)
    fprintf('Parameters for atom %d not found.\n',j);
  end
end
```

```
num_orbitals

% CALCULATION H AND S MATRICES
H_ao = zeros(num_orbitals, num_orbitals);
S_ao = zeros(num_orbitals, num_orbitals);
atomj = 1;
for j = 1:num_orbitals
  fprintf('Interactions for orbital %d\n',j);
  % Determine whether we've already gone through
  % all orbitals on atom;
  if j > sum(orb_list(1,1:atomj))
    atomj = atomj + 1;
  end
  % Set position of nucleus orbital sits on.
  atom_pos1 = atom_pos(atomj,:);

  coeff1 = zeros(1,9);
  % Determine which orbital on atom we are dealing
  % with.
  ao_num = j;
  if (atomj > 1)
    ao_num = j - sum(orb_list(1,1:(atomj-1)));
  end
  coeff1(ao_num) = 1;

  atomk = 1;
  for k = 1:num_orbitals
    if k > sum(orb_list(1,1:atomk))
      atomk = atomk + 1;
    end
    % Set position of nucleus orbital sits on.
    atom_pos2 = atom_pos(atomk,:);
    coeff2 = zeros(1,9);
    % Determine which orbital on atom we are
    % dealing with.
    ao_num = k;
    if (atomk > 1)
      ao_num = k - sum(orb_list(1,1:(atomk-1)));
    end
    coeff2(ao_num) = 1;
    if((j==k)&&(norm(trans_vector) < 0.001))
      if(ao_num == 1)
```

```
      H_ao(j,j) = params(atomj,2);
      S_ao(j,j) = 1;
   end
   if(ao_num>1)&&(ao_num<5)
      H_ao(j,j) = params(atomj,5);
      S_ao(j,j) = 1;
   end
   if(ao_num>4)&&(ao_num<10)
      H_ao(j,j) = params(atomj,8);
      S_ao(j,j) = 1;
   end
 else
   if(atomj~=atomk)||(norm(trans_vector) > 0.001)
      % Don't waste time calculating overlaps
      % between orbitals on same atom.
      [H_ao(j,k) S_ao(j,k)] = ...
 STO_Hij_Sij(atom_pos1,coeff1,params(atomj,:),...
   atom_pos2+trans_vector,coeff2,params(atomk,:));
      end
    end
   end
 end
```

With the solution to Exercise 14.2, the code is now in place for calcu-lating the terms $\left\langle \phi_{j,000} \left| H \right| \phi_{k,\Delta n_a \Delta n_b \Delta n_c} \right\rangle$ in our expression for the matrix ele-ments for the interactions between the irreducible representations for the atomic orbitals within the unit cell at **k**:

$$H_{jk}(\mathbf{k}) = \sum_{\Delta n_a} \sum_{\Delta n_b} \sum_{\Delta n_c} e^{-i\mathbf{k}\cdot(\Delta n_a \mathbf{a} + \Delta n_b \mathbf{b} + \Delta n_c \mathbf{c})} \left\langle \phi_{j,000} \left| \hat{H} \right| \phi_{k,\Delta n_a \Delta n_b \Delta n_c} \right\rangle$$

All that remains for creating the full Hamiltonian at **k** is to apply the phase factors and carry out the sums over Δn_a, Δn_b, and Δn_c. To do this, we will create another function, build_hamiltonian_k. In terms of calcu-lating the phase factors, let's first simplify the dot product in a way that doesn't require us to explicitly determine the reciprocal lattice vectors:

$$\mathbf{k}\cdot(\Delta n_a \mathbf{a} + \Delta n_b \mathbf{b} + \Delta n_c \mathbf{c}) = (k_x \mathbf{a}^* + k_y \mathbf{b}^* + k_z \mathbf{c}^*)\cdot(\Delta n_a \mathbf{a} + \Delta n_b \mathbf{b} + \Delta n_c \mathbf{c})$$

$$= k_x \mathbf{a}^* \cdot \Delta n_a \mathbf{a} + k_y \mathbf{b}^* \cdot \Delta n_b \mathbf{b} + k_z \mathbf{c}^* \cdot \Delta n_c \mathbf{c}$$

$$= 2\pi(k_x \Delta n_a + k_y \Delta n_b + k_z \Delta n_c)$$

The (k_x, k_y, k_z) vector then represents **k** in terms of fractions of **a***, **b***, and **c***. We can then use `for` loops in `build_hamiltonian_k` to carry out the sums in the above equation for $H_{jk}(\mathbf{k})$:

```
function [Hk, Sk] = build_hamiltonian_k(H_ao,...
  S_ao,overlaps,k)

overlap_counter=0;
Hk = 0*H_ao(:,:,1);
Sk = 0*S_ao(:,:,1);
for j1=-overlaps(1):overlaps(1)
  for j2=-overlaps(2):overlaps(2)
    for j3=-overlaps(3):overlaps(3)
      overlap_counter=overlap_counter+1;
      Hk = Hk + exp(-i*2*pi*((k(1)*j1+k(2)*j2+k(3)...
        *j3)))*H_ao(:,:,overlap_counter);
      Sk = Sk + exp(-i*2*pi*((k(1)*j1+k(2)*j2+k(3)...
        *j3)))*S_ao(:,:,overlap_counter);
    end
  end
end

Hk = (Hk+Hk')/2;
Sk = (Sk+Sk')/2;
```

The last two lines of the function are added as a quick guarantee that the Hamiltonian and Overlap matrices stay Hermitian ($H_{ij} = H_{ji}^*$, and $S_{ij} = S_{ji}^*$), in the face of any possible rounding errors that might are arise in the numerical work.

EXAMPLE: THE BAND STRUCTURE OF GRAPHENE

Let's now see how this code can be applied to carrying out a band structure calculation on a model system, a graphene sheet. We already described above how the geometry for this structure can be specified with the `–geo` and `–cell` files; for this we set the base filename as `filename = 'graphene'`. We should then decide on some parameters for the calculation. First is the number of overlaps to consider. If we are primarily interested in the nearest-neighbor C-C interactions, there is no need to calculate interatomic interactions for $|\Delta n_a|$ and $|\Delta n_b| > 1$, while no overlaps need to be considered along the **c** repeat vector, as we are focusing on an isolated layer. This is entered with the line: `overlaps = [1 1 0]`.

Some other parameters we should consider are the *k*-points that we would like to include. There are two ways that we can sample *k*-space. The first is to investigate how the band energies evolve as a function of **k**: it is traditional to select a series of high symmetry points in the Brillouin Zone (see Bradley, C. J.; Cracknell, A. P. *The Mathematical Theory of Symmetry in Solids*; Clarendon Press: Oxford, 2010) and calculate band energies for paths connecting these points. For graphene, we choose the special points as follows, introducing a new variable (nspecial_points) to keep track of the number of special points:

```
kspecial_points = [0.0  0.50  0.0    % M point
                   0.0  0.00  0.0    % Gamma point
                  -1/3  2/3   0.0    % K point
                   0.0  0.50  0.0 ]; % M point

  nspecial_points = size(kspecial_points,1);
```

We can then break up the paths between the points into a series of steps (nsteps) and compile a list of *k*-points (klist) for the full tour of the Brillouin Zone:

```
kcounter=0;
nsteps = 40;
for j1=1:nspecial_points-1
  for step=0:nsteps-1
    kcounter=kcounter+1;
    klist(kcounter,1:3) = (1-(step/nsteps))*...
      kspecial_points(j1,:) + ...
      (step/nsteps)*kspecial_points(j1+1,:);
  end
end
kcounter=kcounter+1;
klist(kcounter,1:3)=kspecial_points...
  (nspecial_points,:);
```

With that, we are now ready to perform the actual calculations. We start by calculating our table of interatomic H_{ij} matrix elements:

```
[H_ao, S_ao, orb_list] = ...
  build_hamiltonian_trans(filename, overlaps);
```

Then we set up an array to collect the band energies for the *k*-points:

```
bandE = zeros(sum(orb_list),kcounter);
```

And next we loop over the *k*-points, calculating the Hamiltonian matrix for each point and diagonalizing it to get eigenvalues:

```
for k = 1:kcounter
  [Hk, Sk] = build_hamiltonian_k(H_ao,S_ao,...
    overlaps,klist(k,:));
  bandE(:,k) = eig(Hk);
end
```

Once this loop is finished, we have the full set of band energies, and turn our attention to plotting them. The most straightforward way is to go band-by-band and plot their energies, with energy along the *y*-axis and our steps through *k*-space along *x*:

```
figure;
hold on;
nbands = size(bandE,1);
for band=1:nbands
  plot(bandE(band,:),'color',[0 0 0]);
end
```

To make the graph more readable, we then add in vertical lines at the special points, and adjust the annotations:

```
kcounter=0;
for j1=1:nspecial_points
  for step=0:nsteps-1
    if step == 0
      plot([kcounter+1,kcounter+1],[-30 10],...
      'color',[0 0 0]);
    end
    kcounter=kcounter+1;
  end
end

xticks([1 41 81 121]);
xticklabels({'M', '\Gamma', 'K', 'M'});
ylabel('Energy (eV)');
axis([1 121 -30 10]);
```

By following these steps, you should obtain a band structure diagram similar to that in Figure 14.2. Here, the eight atomic orbitals per unit cell have given rise to eight bands that move up and down with changes in **k**. This graph makes this **k**-dependence of the band energies very clear.

DETERMINING THE FERMI ENERGY FOR GRAPHENE

However, it does not represent a systematic sampling of the states available to electrons throughout the Brillouin Zone. For example, we cannot judge from this graph where the **Fermi energy** (E_F) separating filled and empty states will lie, and thus which bands are filled, partially filled or empty. To obtain the E_F (and other average properties for the system), we must instead calculate the band energies and wavefunctions for a regular

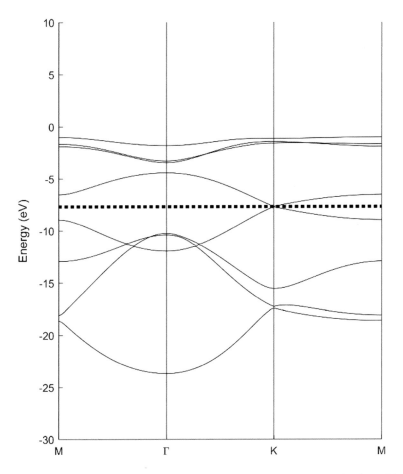

FIGURE 14.2   Simple Hückel band structure for a graphene sheet. The dotted line represents the Fermi energy.

grid of *k*-points rather than linear paths—the finer the grid, the better we capture the continuous nature of the energy bands. To create this grid and build up the band energies for each *k*-point, we simply carry out for loops for k_x and k_y:

```
step = 0.01;
kcounter2 = 0;

for kx = -0.5:step:(0.5-step);
  for ky = -0.5:step:(0.5-step);
    kcounter2=kcounter2+1;
    klist(kcounter2,1:3)=[kx ky 0.0];
    [Hk, Sk] = build_hamiltonian_k(H_ao,S_ao,...
      overlaps,klist(kcounter2,:));
    bandE_grid(:,kcounter2) = eig(Hk);
  end
end
```

Here kcounter2 corresponds to the total number of *k*-points in our grid, which is equivalent to considering a surfaceless crystal containing kcounter2 unit cells. With each *k*-point having 8 band energies, we then have a total of kcounter2*8 band energies in our calculation. As each unit cell contains two carbon atoms, it contributes 4+4 = 8 electrons to the band structure, for a total of kcounter2*8 electrons to populate the bands.

Following these considerations, the E_F can be obtained by sorting the band energies from lowest to highest, and starting from the bottom putting two electrons in each state until we run out:

```
bandlist = sort(reshape(bandE_grid,nbands*...
kcounter2,1));
bandoccups = 0*bandlist;
nstates_total = nbands*kcounter2;
nelectrons_total = 8*kcounter2;
nelectrons_left = 8*kcounter2;

nfilled_bands = 0;
while (nelectrons_left > 0)
  if(nelectrons_left > 2)
     nfilled_bands = nfilled_bands+1;
     bandoccups(nfilled_bands,1) = 2;
     nelectrons_left=nelectrons_left-2;
```

```
   elseif(nelectrons_left > 0)
      nfilled_bands = nfilled_bands+1;
      bandoccups(nfilled_bands,1) = ...
          nelectrons_left;
      nelectrons_left=0;
   end
end

if(bandoccups(nfilled_bands,1) < 2)
      FermiE = bandlist(nfilled_bands,1);
   end
if(bandoccups(nfilled_bands,1) == 2)
      FermiE = (bandlist(nfilled_bands,1)+ ...
                bandlist(nfilled_bands+1,1))/2;
   end
fprintf('Fermi energy = %f\n',FermiE);
plot([1,kcounter],[FermiE FermiE],':');
```

Where the last line plots the E_F as a dotted line on our band structure.

Exercise 14.3. **Calculating the E_F of graphene.** Carry out the above operations and show that the E_F obtained agrees with that shown in Figure 14.2. Note that this lies just as the meeting point of two bands at the special point labeled K. This coincidence of the E_F with the contact point of two bands makes graphene of particular interest for the field of quantum materials.

Exercise 14.4. **Definition of the E_F.** Strictly speaking, the E_F is defined as the chemical potential for electrons in the system: $E_F = \partial E / \partial N_{electrons}$. For a closed-shell molecule, this reduces to simply the average of the HOMO and LUMO energies, i.e. $E_F = (E_{HOMO} + E_{LUMO})/2$. Explain how this definition is implemented in the MATLAB code above.

Index

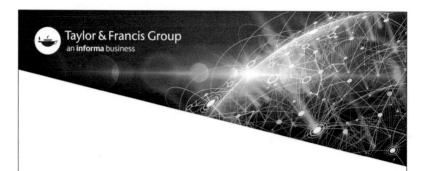